The Description of Nature

The Description of Nature

Niels Bohr and the Philosophy of Quantum Physics

JOHN HONNER

CLARENDON PRESS · OXFORD
1987

Oxford University Press, Walton Street, Oxford OX2 6DP
Oxford New York Toronto
Delhi Bombay Calcutta Madras Karachi
Petaling Jaya Singapore Hong Kong Tokyo
Nairobi Dar es Salaam Cape Town
Melbourne Auckland
and associated companies in
Beirut Berlin Ibadan Nicosia

Oxford is a trade mark of Oxford University Press

Published in the United States
by Oxford University Press, New York

British Library Cataloguing in Publication Data
Honner, John
The description of nature: Niels Bohr
and the philosophy of quantum physics
1. Bohr, Niels 2. Nuclear physics
I. Title
539.7 QC171.2
ISBN 0-19-824976-4

Library of Congress Cataloging in Publication Data
Honner, John.
The description of nature.
Bibliography: p.
Includes index.
1. Quantum theory. 2. Physics-Philosophy.
3. Philosophy of nature. 4. Bohr, Niels Henrik David,
1885-1962. I. Title.
QC174.12.H663 1987 530.1'2 87-23988
ISBN 0-19-824976-4

Set by Downdell Ltd, Abingdon, Oxon
Printed and bound in Great Britain by
Biddles Ltd, Guildford and King's Lynn

Preface

'Nature to him was an open book,' said Einstein of Newton, 'whose letters he could read without effort.' Niels Bohr, rightly belonging in the company of Newton and Einstein, has been regarded by many as having totally revised the Newtonian book of nature. Others, however, have accused him of turning both Newton's book and nature itself into an ethereal vapour. This present work is a defence of the depths of Bohr's vision against those who would dismiss him as utterly obscure or narrowly positivist.

Bohr reflected much on the proper use of descriptive concepts in unambiguous communication about our observations of atomic events. His conclusions lead to a radical broadening of our inherited frameworks for the description of nature. Bohr dwelt in a world that was open rather than closed.

A general introduction to my purpose and plan is given in the first sections of the opening chapter. Readers who are familiar with the preliminaries may prefer to proceed directly to Chapter 3, which occupies the central part of the text as well as being central to the book. May I also warn that Bohr's thought is notoriously elliptic, and the survey of different aspects of his vision demands what may seem to be an inordinate number of treks across the same ground.

The work here relies heavily on quotations from Bohr's published and unpublished writings. The Niels Bohr Archive, thanks to Bohr's fastidiousness and the foresight of his wife and his assistants, holds a rich collection of various drafts of his notes and letters, unique in the history of physics. Rather than paraphrase his ideas, I thought it was important to let his own sometimes inventive language present his equally original outlook. It should also be noted, however, that Bohr may not have considered 'final' many of the remarks which are quoted here from his manuscripts, and such remarks should not be given the same weight as those from his letters or published papers.

While this work is neither a study of Bohr as a person, nor an evaluation of his many major contributions to physics itself, I have taken the liberty of introducing a number of anecdotes into the text in order to allow the splendid personality of the man to lighten some otherwise ponderous discussions. I hope also that the patterns of argument here will be accessible to those who are not specialists either in physics or philosophy. At the same time, it must be acknowledged that many issues, both philosophical and quantum theoretical, entail more subtleties than are touched on here. This is especially the case with respect to the discussion of Bell's theorem and the debates about positivist and non-positivist views of science.

Several acknowledgements are in order. Above all, I thank Professor Aage Bohr for permission to work in the Niels Bohr Archive and to quote so extensively from his father's notes, letters, and writings. Acknowledgement is also due to Pergamon Journals Ltd. for allowing me to rework material from my article 'The Transcendental Philosophy of Niels Bohr' in *Studies in History and Philosophy of Science* 13 (1982), and to Karl E. Peters, editor of *Zygon: Journal of Religion and Science*, for permission to do the same with 'Niels Bohr and the Mysticism of Nature' in *Zygon* 17 (1982). The figures used in Chapter 4 are taken from Bohr's contribution to P. A. Schilpp (ed.), *Albert Einstein: Philosopher-Scientist* and, in a modified form, from Clauser and Shimony's review paper, 'Bell's Theorem: Experimental Tests and Implications'. Quotations from Einstein's papers and letters are reprinted with permission of the Hebrew University of Jerusalem, Israel.

There are many others to whom a debt of thanks is due. Rom Harré and Edward MacKinnon, neither of whom may agree with all of the opinions expressed here, have been especially helpful at different stages, and I also thank Ron Anderson for several thoughtful discussions of the Bohr-Einstein debates and the implications of Bell's theorem. In particular, I am deeply indebted to Professor Erik Rüdinger for his generous assistance in guiding me through the material in the Niels Bohr Archive and for his advice at various stages of this work, albeit that he may not subscribe entirely to the view of Bohr which I offer. The staff and readers at Oxford University Press also provided much stimulus and encouragement when the task seemed too difficult, and to them also I give thanks. One anonymous reader, in particular, may find many of his/her reflections taken up in this final version.

Lastly, I offer most sincere thanks to the many friends and communities who have supported me through these years of preoccupation with Niels Bohr's world.

Jesuit Theological College, Melbourne
J.H. *March 1987*

Contents

Abbreviations

AHQP Archive for the History of Quantum Physics.

APHK N. Bohr, *Atomic Physics and Human Knowledge* (New York, Wiley, 1958), a collection of essays written between 1933 and 1955.

ATDN N. Bohr, *Atomic Theory and the Description of Nature* (Cambridge, Cambridge University Press, 1934), a collection of essays written between 1925 and 1931.

BSC Niels Bohr Scientific Correspondence, Niels Bohr Archive, with microfilm number.

Essays N. Bohr, *Essays 1958-1962 on Atomic Physics and Human Knowledge* (London, Wiley, 1963).

MSS Niels Bohr Scientific Manuscripts, Niels Bohr Archive, with microfilm number.

NBCW *Niels Bohr Collected Works* (Amsterdam, North-Holland, 1972-), with volume number.

TSAC N. Bohr, *The Theory of Spectra and Atomic Constitution* (Cambridge, Cambridge University Press, 1922), a collection of papers from 1913 to 1921.

1
Actors and Spectators

> We are simultaneously actors as well as spectators on the great stage of life.
>
> Niels Bohr

1.1. *'Like one perpetually Groping'*

Bohr formulated the original quantized model of the atom's electron-shells in 1913 when he was twenty-seven years old. Fundamental to his theory was the introduction of a discontinuity, represented by Planck's constant, into the description of nature. Whereas the classical physics of Newton and Maxwell offered a continuous representation of the mechanisms of the physical world, Bohr's theory implied that areas in the fine structure of matter were not quite so open to total surveillance. This same discontinuity was inherent in the 'new' quantum mechanics fashioned by Werner Heisenberg and others in 1925. Further, Heisenberg's formulation of the indeterminacy relations in 1927 underscored the conclusion that these imprecisions in the description of nature were related not to the limitations of our instruments of observation but to the properties of matter.

In May 1927 Norman Campbell wrote a note in the British journal *Nature* entitled 'Philosophical Foundations of Quantum Theory'. Bohr immediately started work on a reply. At this stage of his life his attentions were almost entirely given to the implications of quantum theory for physics. The final draft of Bohr's reply begins with a sentence which both summarizes his enduring point of view and provides a focus for the work undertaken here: 'Due to the contrast between the principles underlying the ordinary description of natural phenomena and the element of discontinuity characteristic for the quantum theory, we must be prepared that every concept used in accounting for the experimental evidence will have only restricted validity when dealing with atomic phenomena.'[1] As a result of Bohr's

[1] N. Bohr, 'Philosophical Foundations of Quantum Theory', an unpublished manuscript from 1927 reprinted in E. Rüdinger (ed.), *Niels Bohr: Collected Works* 6 (Amsterdam, North-Holland, 1985), 69. This collection is henceforth referred to as *NBCW*.

characteristic thoroughness and preoccupation with detail, culminating in an amusing story about how to sign a letter, two absent-minded professors, lost passports, and a missed train, this carefully worked article was never sent to the publisher. Although at the outset of Bohr's efforts to articulate the implications of the 'new' quantum theory, it none the less establishes his primary concern with how our use of words is tied to our engagement with the world we seek to describe. Bohr constantly sought to clarify and stress the conditions shaping the proper use of classical descriptive concepts in our reports of observations of events in atomic physics.

Reflections on the connections between word and world, between concept and percept, and between language and fact, are usually the stock-in-trade of philosophers. Bohr was, as we shall see, both physicist and philosopher. My interest here is in the way in which his thinking offers us, as he used to put it, 'an epistemological lesson': how our usual framework for the description of nature (constructed on causality, space-time, and strong objectivity) is superseded by the more general framework of complementarity in which subject and object are less sharply separated.

Niels Bohr and Albert Einstein shaped modern awareness of the subtlety of nature. Appropriately, in 1922 they received the Nobel Prize for Physics together, Einstein being belatedly awarded the 1921 prize. Despite this, Bohr has remained partially hidden in Einstein's shadow. Where Einstein was publicly hailed as a visionary, Bohr has been regarded by some as obscurantist, as positivist, as never open to the transcendent. Those who knew Bohr, however, knew better. When he was nearly seventy years of age Einstein wrote:

> That this insecure and contradictory foundation [of physics in the years from 1910 to 1920] was sufficient to enable a man of Bohr's unique instinct and tact to discover the major laws of the spectral lines and the electron shells . . . appeared to me like a miracle—and appears to me as a miracle even today. This is the highest form of musicality in the sphere of thought.[2]

Elsewhere Einstein remarked of Bohr: 'I have full confidence in his way of thinking'; 'He utters his opinions like one perpetually groping and never like one who believes to be in possession of definite truth.'[3]

[2] In P. A. Schilpp (ed.), *Albert Einstein: Philosopher-Scientist* (Library of Living Philosophers 7, Evanston, Ill., Northwestern University Press, 1949), 45-7.

[3] Letters to P. Ehrenfest, 23 Mar. 1922, and to B. Becker, 20 Mar. 1954. See A. Pais, *'Subtle is the Lord . . .': The Science and Life of Albert Einstein* (Oxford University Press, Oxford, 1982), 417.

This book, written a century after Bohr's birth on 7 October 1885, is an exploration of his 'perpetual groping'. My focus shifts from Bohr's physics and the quantum theory of the atom to his philosophical presuppositions, and finally to the instinct and tact which enabled him with such sureness to discover the secret harmony of nature. The task is difficult. Bohr wrote many cryptic essays on atomic physics and human knowledge and, despite his scrupulousness about clarity, his way of thinking has remained too enigmatic for professional philosophers. His interpretation of the implications of quantum mechanics and, in particular, his notion of 'complementarity', remain elusive and open to divergent expositions. This may partly be due to the fact that 'complementarity' itself appears to embrace contrasting positions like idealism and pragmatism. But *pace* Wittgenstein, Bohr would claim that it is as important to stammer about the profound as to speak clearly about the obvious. This is not to say that Bohr was in any way content with vagueness in thought, but that he sought ways to clarify the deeper conditions which shape our descriptions of events at the borders of ordinary experience. At such a point, Bohr and an analytic philosopher might part company. It is my contention, nevertheless, that we not only have much to learn about Bohr, but also much advantage to gain from an exploration of his vision.

Just as epistemology was given new purpose and shape as a result of the rise of classical science, through philosophers like Descartes, Hume, and Kant, so also today a whole new set of factors has to be taken into consideration in any theory of knowledge and reality. Epistemology is once again beholden to the philosophy of science. Although the issues remain the same, the parameters are radically different: if our interest is still in the possibility of true knowledge and the relationship between subject and object, our discussions must take into account new discoveries about the nature of physical reality.

Where Descartes attempted to give vague metaphysics the envied certitude of mathematical science, Bohr did the reverse. He argued that theoretical physics was forced, in the end, to consider some of the epistemological dilemmas which abide in philosophy. He was firmly convinced that the claim to *absolute* objectivity inherent in the ethos of classical physics, in which physical objects are assumed to be grasped in a completely detached and non-subjective matter, was denied to the broader view of science demanded by quantum mechanics. Bohr's account of complementarity offered, as he saw it, the only possible framework for objective descriptions. His position was founded partly on the evidence of physics and the implications of the quantum postulate, and partly on the epistemological argument that it is impossible to make an absolute distinction between subject and object in human knowing. Bohr's reflections on the implications of quantum theory and on the conditions for unambiguous communication are aimed at offering a new account of objectivity and realism.

The paradox of the mutuality of subject and object had captivated Bohr since his youth, when he read Poul Martin Møller's story, *The Adventures of a Danish Student*, in which a scholar, on being reproached for having produced so little, protests:

> I get to think about my own thoughts of the situation in which I find myself. I even think that I think of it, and divide myself into an infinite regressive sequence of 'I's who consider each other. I do not know at which 'I' to stop as the actual, and in the moment I stop at one, there is indeed again an 'I' which stops at it. I become confused and feel a dizziness as if I were looking down into a bottomless abyss, and my ponderings result finally in a terrible headache.[4]

This story was almost compulsory reading for those who studied with Bohr at his institute in Copenhagen. In his writings, also, he frequently appealed to the Eastern aphorism: 'We are both actors and spectators on the stage of life.'[5] Herein lie the seeds of the notion of complementarity.

Pascual Jordan, one of the artisans of the matrix formalism for the 'new' quantum theory of 1925, was typical of the many young physicists Bohr took under his wing in Copenhagen. In a moving epitaph, Jordan offers a glimpse of Bohr's vision and stature:

> Bohr had later explained to us in manifold ways how pressingly the notion of complementarity, which had been discovered through quantum physics (as a crystal clear exemplar), raised questions about the relationships between thought and reality. These questions confront us intrinsically everywhere in our struggles to achieve an intellectual grasp of the world. This is surely doubtless, since 'we are simultaneously actors as well as spectators on the great stage of life'.
>
> The distressing, unexpected news that Niels Bohr had on 18 November 1962 departed from this stage leaves all physicists with the feeling that a great epoch

[4] Quoted in N. Bohr, *Essays 1958-1962 on Atomic Physics and Human Knowledge* (London, Wiley, 1963), 13. See also R. Moore, *Niels Bohr: The Man and the Scientist* (London, Hodder & Stoughton, 1967), 15 f.

[5] See N. Bohr, *Atomic Theory and the Description of Nature* (Cambridge, Cambridge University Press, 1934), 119; *Atomic Physics and Human Knowledge* (New York, Wiley, 1958), 20, 63, 81; *Essays 1958-1962 on Atomic Physics and Human Knowledge*, p. 15. These works will henceforth be referred to as *ATDN, APHK*, and *Essays*, respectively. Bohr used this aphorism prior to giving it an Eastern origin: see Kalckar's comments in *NBCW* 6, pp. xxi f.

of physics is over. With his departure we have lost not only one of those men who by their thinking have changed the character of physics: the transformation of our patterns of thought which he brought about took us far—and irretrievably so—from those ordinary ways of looking at things which had developed over two thousand years of human intellectual history and which, in fixed and exclusive form, had come to prosper. Moreover, above all these things, in his person the father of the great world-wide family of quantum physicists had left us.[6]

For Bohr, it was impossible to defend an exhaustive description of reality which presupposed an absolute dividing line between knower and known. Any attempt at exhaustive description, therefore, had to take into account the conditions for distinguishing between subject and object. This was the epistemological conclusion he persistently brandished in the debates over the interpretation of quantum theory and the problems of measurement and observation. In this manner, too, he argued for a new framework for objectivity and exhaustive description. By 'subject', though, Bohr does not mean what philosophers might call the conscious mind, though this is a common enough misreading of his case, but rather the observer *qua* a physical system. We shall return to these arguments later.

The point being made now is that Bohr was, as his most eminent protégé Werner Heisenberg put it, 'primarily a philosopher, not a physicist'.[7] Bohr's philosophy, though, is intuitive and *sui generis*. It is part of our task to locate his enquiry into the deep truths of knowledge and nature in the mainstream of Western philosophy. Many commentators, as we shall see, have placed Bohr in the positivist camp. This seems to me to be a terrible mistake. Others, more understandably, have struggled to situate him at a fixed point on the realist/anti-realist scale. If Bohr is to be pinned down anywhere, then let it be near the descriptive metaphysicians and critical realists.

The convergence of microphysics and metaphysics may come as a shock to some. Classical science, the physics of Newton and Maxwell, has surely outgrown that branch of medieval metaphysics known as natural philosophy, even if vestiges of the old scheme of things remain in the title of Newton's epoch-making *Philosophiae Naturalis Principia Mathematica*. Furthermore, there is great danger in putting new wine in

[6] P. Jordan, *Der Naturwissenschaftler vor der religiösen Frage* (Oldenburg and Hamburg, Stalling, 1972), 214.

[7] W. Heisenberg, 'Quantum Theory and its Interpretation', in S. Rozental (ed.), *Niels Bohr: His Life and Work* (Amsterdam, North-Holland, 1967), 95.

old wineskins: one might lose both. It is equally dangerous to put standardized labels on highly idiosyncratic thinkers. Bohr's vision *is* unique to Bohr. None the less, either for the sake of perspective or out of philosophical possessiveness—everyone loves a champion—Bohr has variously been claimed by this or that school of philosophy: pragmatism, instrumentalism, positivism, relativism, and so on, as we shall see in Chapter 3. His mentors have been identified as Kierkegaard, James, Mach, and Kant. Though there is little warrant for these claims, it still remains important to understand Bohr's own philosophical bent correctly and to situate it in the ebb and flow of human intellectual endeavour. Indeed, it is a conclusion of Bohr's own making that we can only appropriate his vision through our own existing patterns of thought.

Bohr's awareness of the implications of quantum physics forced him to consider the fundamental questions about knowing and objectivity which have concerned philosophers since the beginning of human speculation. It is even possible, perhaps, that the reverse is also true: that his early reflections on the actor-spectator dilemma shaped his formulation of the original quantum postulate in 1913. It is certainly true, I will argue, that microphysics has become, in Bohr's work, a kind of metaphysics.

This is not, then, primarily a historical study. The evolution of the quantum theory and the development of Bohr's interpretation of the quantum formalism have been admirably documented;[8] and Bohr's own collected writings, on physics and philosophy, are now also being made available to us.[9] I do not intend to duplicate or summarize these admirable labours. With the help of such research, however, I want to focus in particular on Bohr's essays on atomic physics and human knowledge. Here his primary concern is with the tension between sense-experience and the subject's inherited framework for communication in any unambiguous description of the material world. Bohr's interest, of course, is in the problem of the observation of sub-quantum

[8] Most eminently by J. Mehra and H. Rechenberg, *The Historical Development of Quantum Theory*, vols. 1-5 (New York, Springer, 1982); J. L. Heilbron and T. S. Kuhn, 'The Genesis of the Bohr Atom', *Historical Studies in the Physical Sciences* 1 (1969), 211-90; M. Jammer, *The Conceptual Development of Quantum Mechanics* (New York, McGraw-Hill, 1966), and *The Philosophy of Quantum Mechanics* (New York, Wiley, 1974); E. MacKinnon, *Scientific Explanation and Atomic Physics* (Chicago, University of Chicago Press, 1982); K. Stolzenburg, *Die Entwicklung des Bohrschen Komplementaritätsgedankens in den Jahren 1924 bis 1929* (Stuttgart, Ph.D. thesis, 1977).

[9] L. Rosenfeld, J. Rud Nielsen, and E. Rüdinger (eds.), *Niels Bohr: Collected Works* 1-11 (Amsterdam, North-Holland, 1972-), later volumes still in preparation.

events. When viewed in another way though, he is constantly attending to the relationship between theory and observation, between word and world, and between language and fact. From this perspective we can examine the proximity of Bohr's position to that of the non-positivist or holist philosophers of science.

Bohr can best be understood through a consideration of the character of his fundamental arguments. The kinds of claims that he makes are, in my view, 'transcendental'. This is not so much to identify Bohr with Kant as to say that he begins his thinking with a reflection on the necessary conditions of the possibility of human experiential knowledge. Or again, he makes assertions about 'what cannot but be the case' in our experience and description of the world. It is in this broad sense that I am using the term 'transcendental', of which more will be said in a moment. If we look at Bohr's work this way, we will come to see why his approach has been regarded by many as elusive, and why he was necessarily and constantly groping after a truth which he knew he could never completely possess. Before we reach this stage, however, some preliminary points must be dealt with in order to clarify later discussions.

1.2. *Outline of Plan: From Easy Truths to Deep Truths*

The structure of this book, as befits its central character, is not without complications. In each of the following chapters I try to isolate particular elements in the compound of Bohr's thought. This means that on some occasions the reader is asked to put aside questions until later, but on others to refer back to earlier discussions. An outline of the plan of the book, therefore, may be helpful here.

The remaining sections of this introductory chapter contain an account of the precursors to my own interpretation of Bohr's thought, followed by some general remarks about recent currents in the philosophy of science in order to situate later appraisals of Bohr's position. I also try to clarify two interpretative terms, 'transcendental' and 'holist', which are of considerable importance in later chapters.

Chapter 2 presents a relatively incontestable account of the genesis of quantum theory and the issues arising out of the new physics with respect to our place in nature. As well as considering the origins of quantum mechanics, the surface features of Bohr's interpretation of the problem of objective measurement are also surveyed.

Focus then shifts in the succeeding chapters to increasingly debatable issues: what I have termed Bohr's transcendental philosophy, his intuition into the harmony of nature, and his metaphysics. Even in the early, more straightforward sections of the text there lurk qualifications which serve as clues to the ensuing investigations. The movement, as Bohr used to say, is from the easy truths to the deep truths:

> In the Institute in Copenhagen . . . we used, when in trouble, often to comfort ourselves with jokes, among them the old saying of the two kinds of truth. To the one kind belong statements so simple and clear that the opposite assertion obviously could not be defended. The other kind, the so-called 'deep truths', are statements in which the opposite also contains deep truth.[10]

The deep truths, of course, are the most interesting ones. They also provide room for contention. Though what is argued in this work may well be challenged by others, my intention is to illustrate the depth of Bohr's thought and its implications. Because the picture of Bohr that is presented here is an uncommon one, the argument is at times laborious only so that it might be less easily refuted. Contrary articulations are perhaps also arguable, but that is of the nature of the deep truths. It is also, as we shall see, of the nature of transcendental claims.

To return to the plan of the book: Chapter 3 directs attention to the underlying arguments—which I describe as transcendental in character—in Bohr's account of quantum physics and his general epistemological description of our activity as subjects in nature. Chapter 4 is devoted to the Bohr-Einstein debates. Here we explore, from a more practical point of view, Bohr's appraisal of realism and objectivity in physics and his insistence on the completeness or universality of the quantum-theoretical description. The disagreement between Bohr and Einstein over the interpretation of the quantum formalism has also frequently been construed as evidence of the diametric opposition of their fundamental outlooks. Assuming the profundity of Einstein's vision, Bohr's could therefore be dismissed as shallow and glib. While commenting on these disagreements, I am also anxious to show the extent of the agreement between Bohr and Einstein in their vision of the subtlety of our place in the universe.

Within the context of recent non-positivist philosophies of science, Chapter 5 contains a further discussion of Bohr's way of doing physics, his particular defence of the realist character of quantum theory, his understanding of the role of symbols and pictures, and his arguments for the mutuality of analysis and synthesis in scientific method. The substance of the previous three chapters is drawn together here, and the foundations are set for the final, more contestable, chapters.

[10] *APHK*, p. 66.

In Chapter 6 I assess the validity of Wolfgang Pauli's assertion that Bohr engaged in a kind of mysticism. While Bohr vigorously denied that he practised 'any mysticism foreign to the spirit of science',[11] I shall argue that his approach to physics was guided by (and in the end justified by) 'an ever richer impression of an eternal and infinite harmony'.[12] Such a sensitivity does not of itself constitute mysticism in the strictest sense of the word, and I am not in any way suggesting that Bohr was the same kind of visionary as Meister Eckhart or Rimbaud, for example. Rather, I want to explore Bohr's marvellous instinct for uncovering the secrets of nature and hence furthering the progress of science. Bohr's intuition into the fine structure of nature was integral to the possibility of his science, not foreign to the spirit of it. Further, his understanding of the mutuality of matter and spirit, also discussed in this chapter, is another clue to his account of realism and, as I have called it, his moderate holism.

The final chapter of this book draws together the various threads of the preceding chapters. I try to show how the transcendental approach, moderate holism, and 'relative' realism all belong together: they are alternative formulations of Bohr's fundamental insight that we are simultaneously actors and spectators in our descriptions of the world we inhabit, the world our words constitute, the world that sustains our words. In establishing this overview of Bohr's position, I also place him in the company of the eminent descriptive metaphysicians, Aristotle and Kant. This is not to identify Bohr's thought with the philosophy of either Aristotle or Kant, but rather to indicate the similarities in their efforts to maintain a unity in knowing and yet respect the apparently distinct contributions of the experienced world and the knowing subject. At the very end of this chapter we return to Bohr's claims about the quantum condition, the framework of complementarity, and his epistemological lesson. Though the quantum postulate has not altered metaphysics, its discovery has drawn physics and metaphysics into a closer, and possibly fruitful, proximity.

1.3. *The Appeal to 'Transcendental'*

Quantum mechanics seemed to provide an entirely new theoretical framework for connecting experimental observations. An indeterminate interplay between the observed system and the observing mechanism

[11] Ibid., p. 91.

[12] N. Bohr, Address given in Copenhagen, 21 Sept. 1928. For the context of this remark, see section 6.2. below.

now became of great importance. Bohr's interpretation of the implications of the new theory, therefore, rested, among other things, on the conditions which made unambiguous reporting of experimental observation possible.

Inasmuch as the quantum theory dealt with experimental observation of classically observed 'properties', such as position or momentum, rather than of the 'thing in itself', any subsequent philosophical reflections were likely to interpret the theory in the light of Kant's critical philosophy or, rather differently, of one version or another of Machian positivism. In this section I would like to make some comments on the sort of arguments which Kant called 'transcendental'. In the following section I will return to the more positivistic interpretations of Bohr's work.

The identification of Bohr's philosophical position which I am proposing here is not without precedent. In 1937 Grete Hermann and her collaborators presented a study of the implications of modern physics for epistemology which contained several references to the Kantian resolution of the perplexity of human knowledge and to transcendental arguments.[13] Later, C. F. von Weisäcker, who as a young man had worked with Heisenberg and Hermann, constantly insisted that transcendental arguments were essential to a defence of Bohr's position. He does not, however, identify such arguments in Bohr's own writings: 'Bohr was essentially right', says Weizsäcker, 'but . . . even he himself did not know why.'[14] He further observes:

> Niels Bohr is the only physicist in our time who—as far as I know, without having been influenced by Kant—proceeded from fundamental insights similar to Kant's . . .
>
> I join Kant in the conjecture that these principles, to put it in Kant's terminology, will be neither transcendent nor empirical, but transcendental in nature. In other words, they will formulate neither metaphysical hypotheses nor particular experiences, but merely the preconditions of the possibility of experience as such.
>
> Only in this framework will physicists be able to do justice to Bohr's doctrine of the indispensability of classical concepts.[15]

Again, in another discussion, Weizsäcker sets out to defend Bohr with 'arguments which Kant called transcendental'.[16] It is my contention

[13] G. Hermann, E. May, Th. Vogel, *Die Bedeutung der modernen Physik für die Theorie der Erkenntnis* (Leipzig, Hirzel, 1937), 44, 78, 136, 192.

[14] C. F. von Weizsäcker, *The Unity of Nature* (New York, Farrar Strauss Giroux, 1980), 186; see also p. 252.

[15] Ibid., pp. 342 f., 345.

[16] Weizsäcker in E. W. Bastin (ed.), *Quantum Theory and Beyond* (Cambridge, Cambridge University Press, 1971), 327.

that Bohr in fact provided himself with the same defence, albeit a somewhat obscured one. Bohr's 'indispensability claims' are, as we shall see in the discussion of transcendental arguments, precisely equivalent to the manner of approach outlined by Kant. In other words, Bohr tries to articulate that which is indispensably the case in any report of human experiential knowing.

Attempts to link Bohr with Kant have been made before, but the majority of them are concerned with Bohr's use of the term 'phenomena' rather than with his underlying method.[17] Weizsäcker is not alone, though, in his more penetrating identification of the similarities between Bohr and Kant in style of argument. Edward MacKinnon, for example, chooses to refer to P. F. Strawson's descriptive metaphysics rather than to Kant's, but he also acknowledges the transcendental character of Bohr's own approach: 'Bohr never intended or attempted a general doctrine of meaning. His problem was the more limited one of clarifying the conditions of the possibility of unambiguous communication. This is a transcendental deduction in the sense that one goes from a fact . . . to the conditions of possibility of this fact.'[18]

Paul Feyerabend also argued at one time that if Bohr was a positivist, then his position was best described as 'higher order' positivism. Elsewhere Feyerabend noted that Bohr's interpretation of quantum physics displays an implicit 'transcendental induction'.[19] In a more negative appraisal, Stephan Körner also labelled Bohr as a transcendentalist, but without further elaboration.[20]

More to the point here is the work of Henry Folse. On the one hand he admits that Bohr's central argument rests on a consideration of the conditions of the possibility of unambiguous communication, but he then observes:

> this . . . has a certain Kant-like appearance, for it issues in statements about the logical requirements for the proper use of concepts in experience. But such an appearance is revealed to be mistaken when we recall that these claims have nothing to do with how experience is obtained, as the Kantian statements

[17] See, for example, C. A. Hooker, 'The Nature of Quantum-Mechanical Reality', in R. G. Colodny (ed.), *Paradigms and Paradoxes* (Pittsburgh, University of Pittsburgh Series in the Philosophy of Science 5, 1972), 168 ff.

[18] MacKinnon, *Scientific Explanation and Atomic Physics*, p. 354.

[19] P. Feyerabend, 'Complementarity', *Aristotelian Society Supplement* 32 (1958), 82, and *Realism, Rationalism and Scientific Method* (Cambridge, Cambridge University Press, 1981), 222.

[20] S. Körner, 'Transcendental Tendencies in Recent Philosophy', *Journal of Philosophy* 63 (1966), 557 f.

concerning the proper use of concepts imply, but only has to do with the communication of an observation or an experience once that experience has already come to pass. In this sense, Bohr's reference to the use of concepts is tellingly non-Kantian.[21]

I do not disagree with Folse here and I also do not wish to push the comparison between Bohr and Kant too far: both minds are too subtle for that. But Kant did elaborate a style of argument which is now termed 'transcendental', although it may not necessarily be used in precisely the same way in which Kant applied it. And it is this style of argument which provides the key to Bohr's mind.

Kant and Bohr are both concerned with what cannot but be the case. For Kant this necessity is not an empirical demand, though it is connected with the necessary conditions for the possibility of experiential knowledge. For Bohr the necessity is both empirical, inasmuch as it rests on the quantum postulate, and epistemological, given his elaboration of the necessary conditions for unambiguous communication about experience. Kant gave the term 'transcendental' a specific, albeit elusive, meaning. He made a sharp distinction between that which lies entirely outside the sensible world, which he called 'transcendent', and that which in our reasoning refers to the 'totality of conditions in the sensible world', the 'transcendental'.[22]

Behind Kant's definition of 'transcendental' lay what he considered to be a new manner of argument. Prior to Kant, philosophers had either rejected any a priori descending method, as Bacon and Hume had done, or had failed to keep their metaphysical feet on empirical ground, as in the case of Kant's predecessor Christian Wolff, and had literally 'gone over the top' with deductive arguments based on theoretical axioms. Recognizing that on the one hand strict empiricism would ring the death-knell of metaphysics, and that on the other hand pure a priori arguments lacked empirical starting-points, Kant turned to his transcendental approach. He suggested that we begin with the givenness of experience and then, by reflecting on the necessary conditions for the possibility of experience, move to universal conclusions.

A more detailed statement of Kant's position would provoke prolonged debate. Some will argue that his philosophy is essentially idealist, whereas others may claim it to be more even-handed and to

[21] H. J. Folse, 'Kantian Aspects of Complementarity', *Kant-Studien* 69 (1978), 61. Folse has also recently produced *The Philosophy of Niels Bohr: The Framework of Complementarity* (Amsterdam, North-Holland, 1985).

[22] See I. Kant, *Critique of Pure Reason*, A 565/ B 593, A 11-12/ B 25.

entail a moderate realism. Some will take a very specific view of 'transcendental' and restrict it to Kant's interest in the *concepts of understanding* constitutive of experiential knowledge. Others, again, would adopt a broader view and see the term embracing reflections on the conditions of the possibility of experiential knowledge—though this would be taken to mean more than just conceptual explication. It is the latter and broader usage of the term that I wish to adopt in the investigation of Bohr's fundamental arguments.

Strawson has described Kant's interests as 'the investigation of that limiting framework of ideas and principles the use and application of which are essential to empirical knowledge, and which are implicit in any coherent conception of experience which we can form'.[23] This interest also lay close to the centre of Bohr's reflections, for his epistemological lesson began with the conditions of the possibility of unambiguous communication about events on the boundary of our experience of nature. This concern may be 'tellingly non-Kantian', as Folse asserts, in as much as it does not include Kant's focus on the concepts of reason. At the same time, it belongs squarely in the broader application of the term 'transcendental', as used by Kant and by more recent philosophers.

I am therefore using 'transcendental' to signify a concern with the necessary conditions of the possibility of experiential knowledge, which includes unambiguous reports of that knowledge.[24] The term is not in itself of any great significance, and my point is not to identify Bohr with Kant, but rather to describe the character of Bohr's philosophy with reference to this pioneering descriptive metaphysician. Given the history of the term 'transcendental', and its currency in philosophy today, it will serve us well. The term 'transcendental' is a key: an understanding of Bohr lies behind the door that it unlocks.

Much later, in the final chapter of this book, it will be necessary to examine the nature and validity of a transcendental claim. It will suffice here to note that a transcendental argument entails not a syllogism but a blunt assertion about what cannot but be the case. We have to get the point for ourselves by reflecting upon what must be constitutive in any account of our experience. As a very simple example, it is a necessary condition of the possibility of my being able to play chess, or what

[23] P. F. Strawson, *The Bounds of Sense* (London, Methuen, 1966), 18.

[24] See J. R. Honner, 'On the Term "Transcendental" ', *Milltown Studies* 11 (1983), 1-24, for a more detailed discussion of this term and an evaluation of the transcendental argument by Charles Taylor, Strawson, and others.

might pass as chess, that I should understand the rules for moving a queen and a knight and so on. As a more elaborate transcendental reflection, it seems constitutive of any sense-experience that it occurs in a context of space and time, 'somewhere' and 'somewhen'. In this sense, transcendental claims are performative, being disclosed in what is constitutive of our abilities. What is left on paper as the conclusion to a transcendental deduction, then, is the articulation of an insight into the conditions governing one's own experiencing and knowing. For any insight there may be a variety of articulations. And the more interesting the claims, the more contested these articulations will become.

Bohr's framework of 'complementarity' is perhaps the most contested of recent transcendental claims. He asserts, as we shall see, that unambiguous reports on atomic observations require the use of ordinary everyday concepts, drawn from a Newtonian world of space-time and causality, a world of discrete, identifiable objects. Hence we are caught in the position of having to use terms from one view of nature to report on a quite contrary vision, in which sharp subject-object separation and continuity of observation are no longer tenable. What is observed as a wave is also observed as a particle, and the application of mutually exclusive concepts is justified once the framework of complementarity is substituted for that of continuity and univocity. If we are able to understand how Bohr arrives at this prescription, however, we will be half-way towards understanding what he is trying to convey.

Another interpretative term I wish to use in this study is 'holism'. In the section that follows I introduce the term in the context of recent debates between positivist and non-positivist philosophers of science, with more than a passing reference to positivist claims on Bohr's philosophy.

1.4. *Positivism and Holism*

Bohr's philosophy has been judged by some, most notably by Karl Popper and Mario Bunge, to be positivist in character.[25] The positivist philosophy of science, in the guise most closely associated with the interpretation of quantum theory, was initiated by Ernst Mach, though the extent to which Mach himself was a positivist remains open to

[25] See M. Bunge and K. Popper in M. Bunge (ed.), *Quantum Theory and Reality*, Studies in the Foundations, Methodology, and Philosophy of Science 2 (New York, Springer, 1967), *passim*; see also P. Heelan, *Quantum Mechanics and Objectivity* (The Hague, Nijhoff, 1965), p. ix, and Hooker, 'The Nature of Quantum-Mechanical Reality', p. 193.

debate. While it admits of many meanings, positivism is generally taken to include positions holding that only those knowledge-claims which are related directly to experience are valid. One formulation of positivism for example, the instrumentalist philosophy of science, reduces science to the ordering of phenomena or observables. In this phenomenalist scheme, science has little or nothing to do with the unseen nature of things. Instrumentalism is at best very weakly realist. Bohr, given his focus on the observation of quantum processes rather than on quantum objects themselves, is said to belong in such a category. Such a claim, however, can be shown to be of limited validity.

It remains a fact that Bohr regarded science as being concerned with the collecting and ordering of experience; and it is equally obvious that he placed great stress on the essential role of observables in any unambiguous description of nature. In Bohr's mind, on the other hand, the interaction between the means of observation and the event being observed provided the context for a new mode of objective description. Hence his stress on the importance of paying heed to the particular conditions surrounding any individual observation. But Bohr's position is not a subjectivist one, in which it might be claimed that the reality does not exist unless the observer is engaged with it. His entire argument is aimed at providing a framework for applying our limited observation-language to real events lying at the boundary of the univocal application of such concepts. In this manner he moves to 'higher order' positivism, as Feyerabend put it, by way of a transcendental argument. More will be said about this in the chapters to follow.

Weizsäcker has been unrelenting in his defence of Bohr against the charge of positivism. At the outset of his reflections on what he calls the Copenhagen interpretation, he states unequivocally that '*Bohr was no positivist*'. He later explains:

> The fact that classical physics breaks down on the quantum level means that we cannot describe atoms as 'little things'. This does not seem to be very far from Mach's view that we should not invent 'things' behind the phenomena. But Bohr differs from Mach in maintaining that 'phenomena' are always 'phenomena involving things', because otherwise the phenomena would not admit of the objectification without which there can be no science of them. For Bohr, the true role of things is that they are not 'behind' but 'in' the phenomena.[26]

Nevertheless, the appeal of Mach's position becomes more obvious if one calls to mind the difficulties faced by atomic physicists in the early

[26] Weizsäcker, *The Unity of Nature*, p. 185.

1920s, burdened by more and more paradoxical experimental results which they could neither explain nor, more pressing still, mathematically describe. This was particularly the case with observations of the emission spectra of excited atoms: atomic electrons, after being excited to higher energy levels, dropped back to their ground states and emitted energy in the form of highly specific wavelengths of electromagnetic radiation. The usual way of trying to explain these emissions involved taking into account the masses of the electrons, their positions, and the transitions between different orbits. However, these factors seemed to be barely observable in themselves. Heisenberg therefore gave his attention only to quantities which could be found directly in experimental observation, namely the wavelengths and amplitudes of the emitted radiation. His consideration of tables and paired results led to the matrix formulation of the 'new' quantum mechanics.

In 1927 Heisenberg enunciated a fundamental principle which showed close similarities to Mach's philosophy of science. 'Physics', Heisenberg proposed, 'must be confined to a formal description of the relation between perceptions.'[27] Both Bohr and Einstein, however, were very negative in their comments on Heisenberg's proposal. Einstein at this stage had rejected not only Mach's philosophy of science, but also positivism in all its moods and tenses.[28] Heisenberg, too, was later critical of the positivists' foreshortening of the horizon of human enquiry:

> The positivists have a simple solution: the world must be divided into that which we can say clearly and the rest, which we had better pass over in silence. But can any one conceive of a more pointless philosophy, seeing that what we can say clearly amounts to next to nothing. If we omitted all that is unclear, we would probably be left with completely uninteresting and trivial tautologies.[29]

Heisenberg also recalls Bohr commenting on positivism in 1952 as follows:

> I can readily agree with the positivists about the things they want, but not about the things they reject. . . . Positivist insistence on conceptual clarity is, of course, something I fully endorse, but their prohibition of any discussion of the wider issues, simply because we lack clear-cut enough concepts in this realm,

[27] W. Heisenberg, 'Über den anschaulichen Inhalt der quantentheoretischen Kinematik und Mechanik', *Zeitschrift für Physik* 43 (1927), 197.

[28] See A. Einstein, 'Autobiographical Notes', in Schilpp, *Albert Einstein*, p. 24.

[29] W. Heisenberg, *Physics and Beyond*, (London, Allen and Unwin, 1971), 213.

does not seem very useful to me—this same ban would prevent our understanding of quantum theory.[30]

In recent years, philosophers of science have explored the relationships between theory and observation, as well as between subject and object, producing a view of science quite different from that of the positivists. Theory is no longer seen as reducible to experience, given that the language of observation depends upon pre-existing theoretical frameworks; and yet experience, in the form of new experimental evidence, is seen to shape theory.[31] Further, Bohr can be shown to be much more at home among this company than with the positivist collective. In other words, despite his remarks about phenomena and the crucial role of experience, Bohr is just as much concerned with the *mutuality* of observable and observer in the whole process of observation. Through such considerations he justifies the complementary application of otherwise contradictory concepts for an exhaustive account of observations at the quantum level. This aspect of his thought will be referred to as moderate holism.

Moderate holists argue for the possibility of scientific thought engaging a real world. In this realist account truth is taken to entail the asymptotic convergence of theory and reality. Theory is seen as partly entrenched in the real world and, under constant revision, coming closer and closer to it. At the same time, theory and observed reality are not regarded as entirely distinct, but as mutually shaping and hence mutually convergent. Finally, the notion of a universal scientific language has had to be abandoned, because all language is believed ultimately to be related to its practitioners and hence cannot be reduced in any other way.[32]

One consequence of the changes in the philosophy of science has been the shift in emphasis in epistemology from attempts at clear-cut and foundationalist accounts of knowing to restatements of the intangible obscurities encountered in epistemology. This has led to more anarchic, non-positivist views of science, in terms of its objectivity, which can be described as thoroughgoing holism or thoroughgoing relativism.

Richard Rorty, for example, has enchantingly illustrated a connection between foundational epistemologies and positivist philosophies of

[30] Heisenberg, *Physics and Beyond*, pp. 208 f.

[31] I am thinking of the writings of Polanyi, Lakatos, Feyerabend, Kuhn, Harré, Hacking, Hesse, Bhaskar, Dretske, and many others.

[32] See, for example, M. Hesse's more detailed treatment of these issues in her *Revolutions and Reconstructions in the Philosophy of Science* (Brighton, Harvester, 1980), pp. vii ff.

science. He rejects the distinctions between inductive and deductive methods and between analytic and synthetic arguments. Objective knowledge is thus reduced to the conformity of any proposition to the norms of justification. He concludes 'that the notion of knowledge as the assemblage of accurate representations is optional—that it may be replaced by a pragmatist conception of knowledge which eliminates the Greek contrast between contemplation and action, between representing the world and coping with it.'[33] Rorty's pragmatist theory of knowledge, itself a debunking of all theories of knowledge, results in a thoroughgoing holism and provides a strong and engaging position with which to compare and contrast Bohr's more moderate holism. After all, Bohr is concerned with the relationship between 'picturing' reality and reality itself. He rejects naïve realism and a stringent notion of objectivity, yet his holism is moderated by a sense that we can know what we are talking about if we observe the conditions for unambiguous communication. Bohr searches out a new defence of realism.

Like 'transcendental', the term 'holism' plays a prominent part in the explication of Bohr's philosophy given here. It is particularly important in the discussion of his notion of complementarity, a principle which prescribes the necessity of applying mutually exclusive concepts in an exhaustive description of observations at or beyond the bounds of human experience.

Holism is also a term frequently employed in non-positivist philosophies, notably Rorty's restatement of the work of Quine, Sellars, James, Kuhn, and Heidegger, which can help us in our assessment of Bohr's own position. Rorty's vigorous argument provides a useful point of reference, whether for comparison or contrast, for the understanding of Bohr's work. It also constitutes a link with the philosophy of William James, a philosophy which Bohr acknowledged to show similarities to his own outlook, though there is considerable debate about whether Bohr first came across James before 1912 or even as late as 1932.[34] In his last interview with Thomas Kuhn, the day before his death, Bohr recalled his delight in James's way of thinking: 'William James is really wonderful in the way he makes it clear . . . that it is quite impossible to analyze things in terms of—I don't know what one calls them, not atoms—I mean simply if you have some things . . . they are so

[33] R. Rorty, *Philosophy and the Mirror of Nature* (Oxford, Blackwell, 1980), 11.

[34] See Jammer, *The Conceptual Development of Quantum Mechanics*, pp. 176 ff.; K. M. Meyer-Abich, *Korrespondenz, Individualität und Komplementarität* (Wiesbaden, Steiner, 1965), *passim*; and Folse, *The Philosophy of Niels Bohr*, pp. 49 f.

connected that if you try to separate them from each other it has nothing to do with the actual situation.'[35] As we learn more of Bohr's fundamental epistemological convictions we shall see how in many ways these comments about James's holism match Bohr's own outlook.

Rorty attacks the notion that there are 'foundations' for knowledge and that we can provide an architectonic philosophy of science. Like Bohr, he is concerned with mutuality of actor and spectator. Rorty makes his case indirectly, however, by criticizing the idea that knowledge consists in language accurately representing nature. In his thoroughgoing holism, knowledge is not justified by a special relationship between words and objects, but simply by conversational practice. In an echo of Bohr's comments on James, Rorty says:

> We will not be able to isolate basic elements except on the basis of a prior knowledge of the whole fabric within which these elements occur. Thus we will not be able to substitute the notion of 'accurate representation' (element-by-element) for that of successful accomplishment of a practice. . . . This holist line of argument says that we shall never be able to avoid the 'hermeneutic circle'—the fact that we cannot understand the parts . . . unless we know something about how the whole works, whereas we cannot get a grasp on how the whole works until we have some understanding of its parts.[36]

Thoroughgoing holism entails withdrawal from the quest for certainty in knowledge. Rorty freely admits that this is a paradoxical and relativist position, but he does his best to persuade us that it is a wiser and more honest way of viewing knowledge than the schemes proposed in foundationalist epistemologies.

Rorty's work is not only a defence of thoroughgoing holism, but also an attack on the idea that knowledge consists in the internal mirroring of external reality. For Rorty stringent objectivity is impossible, if by objectivity we mean a one-to-one correspondence between object and its conceptualization, since sharp inner-outer distinctions cannot legitimately be defended. Here, too, Bohr would agree with Rorty, although he would advocate a different kind of objectivity. Chapter 5, in particular, provides a more detailed account of Bohr's philosophy of physics and the roles of 'pictures' and symbols in the theoretical representation of nature.

[35] Bohr in *Archive for the History of Quantum Physics* (henceforth *AHQP*, referring to holdings in the Archive for the History of Quantum Physics), last interview, 17 Nov. 1962, p. 5 of transcript.

[36] Rorty, *Philosophy and the Mirror of Nature*, p. 319.

Note also that our usage of holism is not to be confused with that adopted in biological discussions earlier this century. It is not to be interpreted as signifying that everything is part of a single whole and that the parts are thereby of lesser significance. Rather, it connotes the mutuality, or circularity, of the elements in our knowing: word and world, language and fact, concept and percept, and so on. Bohr's framework of complementarity encompasses similar mutualities. His is a moderate holism, however, in as much as he also wants to insist on a practical distinction between concept and reality as a necessary move in the establishment of revised notions of objectivity and realism.

Rorty himself explores the difference between his own thoroughgoing holism and a more moderate view, in which some form of truthfulness to reality is maintained. He rejects, for example, Jürgen Habermas's efforts to provide a transcendental defence for moderate realism. Both Habermas and Rorty attack correspondence-theories of truth, theories in which truth is seen to consist in statements accurately representing what is the case. But where Rorty's heroes debunk any possibility of truthfulness to reality, Habermas provides an explicitly transcendental defence:

> Correspondence-theories of truth tend to hypostatize facts as entities in the world. It is the intention and inner logic of an epistemology reflecting upon the conditions of possible experience as such to uncover the objectivist illusions of such a view. Every form of transcendental philosophy claims to identify the conditions of the objectivity of experience by analyzing the categorical structure of objects of possible experience.[37]

This way of proceeding is, as we shall see, similar to Bohr's investigation of the conditions for unambiguous communication. Bohr strongly rejects the 'objectivist illusion', the notion that the terms of the theory correspond exactly to the system represented by the theory, entailed in the mechanist models inherited from classical science. Likewise, Bohr established a new account of objectivity by exploring the conditions for the unambiguous use of concepts drawn from experience. This corresponds to Habermas's task of 'analyzing the categorical structure of objects of possible experience'. Rorty's 'heroes', on the other hand, 'hammer away at the holistic point that words take their meanings from other words rather than by virtue of their representative character . . . and that vocabularies acquire their privileges from

[37] J. Habermas, Postscript to *Knowledge and Human Interests*, quoted in Rorty, *Philosophy and the Mirror of Nature*, pp. 381 f.

the men who use them rather than from their transparency to the real'.[38] At this point Rorty and Bohr part company. Bohr sees words as taking their meaning both from the world and from other words, and vice versa.

One might suggest to Rorty that his heroes are in fact articulating an insight into what cannot but be the case with respect to our knowing participation in the world. This would be to foist a transcendental philosophy upon them, however, and to put them in a position which they would not wish to occupy: thoroughgoing holism excludes the thought of philosophical underpinnings. It is pragmatic in the extreme.

Bohr, as we will see, does attempt to provide philosophical underpinnings. He describes his conclusions as a primary epistemological lesson. Although he employs what might be termed holist strategies in his defence of the mutuality of conceptual frameworks and experiential knowledge, Bohr's principles of correspondence and complementarity suggest a convergence of experience and theory which is underwritten by his belief in the infinite and eternal harmony of nature. 'We must continually count on the appearance of new facts,' says Bohr, 'the inclusion of which within the compass of our earlier experience may require a revision of our fundamental concepts.'[39] Words may not mirror the world, but word and world are asymptotically convergent.

The holist outlook also pervades Bohr's insistence on the wholeness of quantum-theoretical state-descriptions, a topic which will be dealt with in the discussion of the Bohr-Einstein debates below. Here the peculiar character of the observer/observed interaction at the quantum level throws light on the conditions governing our unambiguous description of nature. In this sense, as Jammer has pointed out, Bohr's work also bears comparison with Aristotle's notion of wholeness in our description of the world.[40]

While Bohr's approach is more epistemological than ontological, his holism is not so thoroughly circular that it prevents him from asserting the possibility of objective descriptions. For him words do more than feed on other words. Because of this conviction, he struggles to establish a new account of objectivity. Again, while he does not argue for a correspondence-theory of truth, neither does he submit to pure pragmatism of meaning and a consistency-theory of truth. He believed that science is moving towards, if never arriving at, the infinite

[38] Rorty, *Philosophy and the Mirror of Nature*, p. 368.
[39] *ATDN*, p. 97; see section 5.5. below.
[40] See Jammer, *The Philosophy of Quantum Mechanics*, pp. 199 f.

coherence and harmony of nature. Thus, while Bohr uses holist arguments about language, his instinct is that communication in terms of observables is more central, more unambiguous, and therefore has a priority in the justification of knowledge. His approach entails a moderate holism and a kind of realism which might be called 'relative'. It is the task of later chapters to clarify this account of realism.

A common thread runs through moderate holism, relative realism, transcendental argument, and descriptive metaphysics. ('Descriptive metaphysics' is used here, as Strawson does, to mean description of the structure of our thought about the world.) Neither dualist nor monist, they all entail an engagement with the complex tensions between knower and known. Precisely because the transcendental method calls for reflection upon the conditions of one's experiential knowledge, concept and reality are intertwined. There is indeed something pragmatic and behaviourist about the transcendental claim, for it is disclosed in the context of performance. Rather than accept the annihilation of epistemology, however, an effort is made to restate and clarify the conditions for knowledge. Whilst it forms part of the circularity in much of our conceptualization of reality, the transcendental path is not viciously circular. As we stalk the perimeter of our questions, so also do we mark the way more clearly for ourselves. 'At any given time', as Hesse puts it, '*some* observation statements result from correctly applying observation terms to empirical situations.'[41]

For the thoroughgoing holist, questions about how language relates to the world cease to be important. So also, the issues of realism and idealism become non-issues. This is not to say that such holism is anti-realist, but that it does away with the question of the correspondence of language to reality altogether. The moderate holist, on the other hand, defends a convergence between language and reality. In this sense moderate holism and a relative realism go together, and I shall argue that Bohr can fruitfully be compared with such a style of philosophy. Subsequently, also, it becomes clear that his position is closer to the non-positivists than the positivists. The following chapters of this book will, I hope, justify the claims made here.

1.5. '*The Only Possible Objective Description*'

Bohr died on 18 November 1962. Only the day before, when interviewed by Kuhn, he lamented the fact that philosophers had failed

[41] Hesse, *Revolutions and Reconstructions*, p. 108. See also D. Papineau's critique of thoroughgoing holism in *Theory and Meaning* (Oxford, Clarendon, 1979), 117-33.

to grasp the central point in his work: 'I think it would be reasonable to say that no man who is called a philosopher really understands what one means by the complementarity description. . . . They did not see that it was an objective description, and that it was the only possible objective description.'[42] Today Bohr's philosophy remains opaque for even some of his most attentive and qualified readers. Abner Shimony, for example, recently stated: 'I must confess that after 25 years of attentive—and even reverent—reading of Bohr, I have not found a consistent and comprehensive framework for the interpretation of quantum mechanics.'[43] If Bohr's thought is not found to provide a consistent framework for the interpretation of quantum mechanics, then perhaps one's expectations of 'interpretation' should be revised. It is central to Bohr's position that the interpretation of the quantum formalism cannot be undertaken as a quest for those distinct items in reality which correspond uniquely to the terms in the theoretical formalism.

This study makes an attempt 'really' to understand Bohr's elusive vision: open rather than constrained, subtle rather than facile, and more preoccupied with the frontiers of human knowing than with the well-worn and familiar. To speak of Bohr as a moderate holist, transcendental philosopher, and descriptive metaphysician may seem outrageous to some, yet the evidence is substantial. Bohr himself would have found it uncomfortable to be assessed within the frameworks I have chosen, but I trust that he would have supported these endeavours in as much as they illuminate the framework of complementarity as 'the only possible objective description'.

Bohr was not a man who was afraid to get out of his depth, though he was cautious about being entangled in academic philosophical debates. He groped for the truth, as Einstein put it, rather than pretending to possess it completely. Dear to Bohr was Schiller's couplet: 'Nur die Fülle führt zur Klarheit, / Und in Abgrund wohnt die Wahrheit',[44] and his principle of complementarity must be anticipated to be both simple and elusive. 'Complementarity', says Rosenfeld, 'is no system, no doctrine with ready-made precepts. There is no via regia to it; no formal definition of it can be found in Bohr's writings.'[45] So, in following

[42] *AHQP*, interview 5, p. 3 of transcript.

[43] A. Shimony, review of Folse's *The Philosophy of Niels Bohr*, in *Physics Today* 38 (1985) 10, 109.

[44] 'Only wholeness leads to clarity, / And truth dwells in the abyss', from Schiller's 'Sayings of Confucius'. See W. Pauli (ed.), *Niels Bohr and the Development of Physics* (London, Pergamon, 1955), 31.

[45] L. Rosenfeld, 'Niels Bohr's Contribution to Epistemology', *Physics Today* 16 (1963) 10, 48.

Bohr's course through the origins of quantum theory and its interpretation, we set out into the deep. Yet, as another commentator on Bohr's world-view put it: 'it is worthwhile to examine a philosophy that has led to concrete physical suggestions . . . that is largely responsible for the discovery and proper understanding of one of the most fascinating contemporary theories and that entails also some very interesting ideas concerning the relation between subject and object, concept and fact, and knowledge and observation.'[46]

Louis de Broglie, a quantum physicist who did much to reconcile the wave and particle theories of light, describes Bohr as 'the Rembrandt of contemporary physics', with a 'predilection for "obscure clarity"'.[47] Something I have stressed in this chapter is that Bohr ought to be given the benefit of 'uncertainty', if not of doubt. Rather than simply obscure, he was constantly seeking the truth, but never claimed to be in complete possession of it. A merciless questioner, he was continually refining his own articulations as much as he was challenging his interlocutors.

Enough for the preliminaries. We will begin our journey with that which can be dealt with most clearly, namely the evolution of quantum physics, and then move to the more obscure, Bohr's interpretation of the implications of the new theory. The last three chapters take up these questions of realism, objectivity, and the connection between language and classical mechanical frameworks in greater detail. First, however, we must consider some physics.

[46] Feyerabend, *Realism, Rationalism and Scientific Method*, p. 248.

[47] L. de Broglie, *New Perspectives in Physics*, English trans. by A. J. Pomerans (New York, Basic Books, 1962), 97 f.

2

Quantum Theory and its Interpretation

> Bohr was convinced right from the start that the stability of Rutherford's model was . . . of 'non-mechanical' origin.
>
> Léon Rosenfeld

2.1. *Models of the Atom*

This chapter contains a brief chronological account of the development of quantum physics in the first three decades of the twentieth century, which provides the background for a discussion of Bohr's initial attempts to interpret the quantum formalism. It also enables us to clarify the meanings which Bohr gives to such key words as 'correspondence', 'complementarity', 'phenomena', 'classical physics', and 'measurement'. In conclusion, we are able to anticipate Einstein's alternative solutions to the puzzles associated with the new theory and to face, with Bohr, questions about the limits to the possibility of experiential knowledge. Although this chapter is devoted to the more obvious surface features of Bohr's thought, and to an introduction to his terminology, Chapter 3 offers a detailed exegesis of his underlying arguments.

Bohr's original quantum theory of the atom was published in 1913. The picture of the atom which Bohr proposed is often introduced in terms of Rutherford's 'planetary' model: the atom is envisaged as a kind of miniature solar system, in which the positively charged nucleus corresponds to the sun, while the negatively charged electrons whirl around the nucleus in the same way that planets orbit around the sun. Planetary orbits are controlled, of course, by a balancing of gravitational attraction and centrifugal acceleration. As the planets slow down, they are drawn closer to the sun.

It is not quite the same with the atom. Here the attractive force is electromagnetic rather than gravitational. Although the atomic orbitals are *observed* to be stable and exhibit regular spectral patterns, Rutherford's model of the atom was in theory mechanically and electromagnetically unstable. Bohr's resolution of these difficulties involved suspending the ubiquity of the usual laws of mechanics. Right from the start, as Rosenfeld comments, Bohr was convinced that observed stability was of non-mechanical origin.[1]

For the sake of those who think they are likely to get lost in the physics here, I would like to digress for a moment: in order to understand Bohr's proposal, I am going to suggest a model of the atom which may convey more to the uninitiated and which can be called the 'bookcase' model. Bohr's original quantum theory of the atom does for electrons more or less what bookshelves and catalogues do for library books: it regulates where electrons should and should not be found; it gives indications of what the various items on the shelves are like, though it does not tell us precisely where they are; and, by and large, it works.

Imagine a pile of books on the floor in front of you: the force of gravity packs them close together in a jumble on the ground; the ordering is more or less random, although the bigger and heavier books will tend to fall to ground level. A similar plum-pudding model of the atom was put forward by J. J. Thomson when charged particles within the atom were first considered: within a positively charged sphere negatively charged electrons were embedded at appropriate distances from each other. In Bohr's atom, following Rutherford's work, the electrons are taken to be in orbits around the positively charged nucleus. But what keeps the electrons in their orbits, and prevents them from repelling each other or from spiralling into the nucleus? Opposite charges do attract, after all, and like charges repel.

Bohr believed that the electrons were located in specific orbits. Stated in terms of our image of a pile of books arranged in a bookcase, this means that the electrons have been given particular shelf locations or orbitals. A sensible bookcase has the biggest shelves near the floor, so that one does not have to use too much energy in moving the heavier books around. In the quantum atom these shelves—the ones which are the first to be filled—are called 'zero' or 'ground-state' energy levels.

In a well-constructed bookcase, the shelves get closer together as the bookcase gets higher: the smaller, lighter, and more easily moved volumes are thus some distance from the ground level. Often the top shelves are left empty, ready for expansion. And if the bookcase were infinitely high, then two things would happen: first, the shelves would become so infinitesimally close together that they would be indistinguishable; and, secondly, the books in them would be so far

[1] See L. Rosenfeld, 'The Wave-Particle Dilemma', in J. Mehra (ed.), *The Physicist's Conception of Nature* (Reidel, Dordrecht, 1973), 253.

removed from the earth's gravitational force that they would be able almost to float free in space. So also with Bohr's atom. The higher shelves correspond to higher quantum states or quantum numbers. And, for such high quantum numbers, the discontinuous steps of the bottom shelves gradually merge into an apparent continuity of possible levels.

Classical physics, constituted by Newton's laws of motion, Maxwell's equations for electromagnetic fields and waves, and statistical thermodynamics, emphasized the *continuity* of motion, fields, and waves. The quantum theory, on the other hand, seemed to some to imply discontinuity at the core of physical reality. The bookcase model illustrates how the great discontinuities in atomic orbitals gradually become indistinguishable on the scale of high quantum numbers, which is also the macroscopic scale of classical physics. This merging of quantum and classical physics, where quantum accounts correspond to classical calculations, will be discussed later in terms of the two versions of Bohr's Correspondence Principle.

Two helpful lessons can be drawn from the bookcase model of electron energy levels in the atom. First, the shelves represent quantum 'stationary states', in as much as they define the limits of stable positioning. For example, one could try to place a book a few centimetres above a particular shelf, but the exercise would be in vain: the force of gravity would immediately drop it to shelf level, a 'stationary state'. Bohr believed that similar limits are imposed upon where orbiting electrons are found. Just as we can move books from one shelf to another, if there is enough space, so also an electron can move from orbit to orbit, *but it cannot occupy any place between these orbits*. The mysterious term 'quantum jump' thus refers to a sudden transition from one level to another.

The second lesson we can learn from the bookcase model has to do with the nature of atomic spectra. A book falling to a stable position on a shelf gives off some sound-energy; even with your back turned to the bookcase, you know that a loud thud indicates the fall of some fat volume or other. In the Bohr atom there are no shelves as such, but in 'falling' from higher to lower orbitals an excited electron gives off a quantity of energy in the form of electromagnetic radiation. Because the orbital levels have been defined, and because the energy given off corresponds to the amount of work required to lift the electron to a higher level, these radiations have very specific and recurring energies (or wavelengths). By examining the intensities and wavelengths of atomic spectra, therefore, physicists are able to gather information about the characteristic energies of the electron orbitals of the atom concerned.

Keeping this model of the quantum atom in mind, we will now examine the genesis of Bohr's theory and the problems it raised: is radiation continuous or discontinuous; is classical physics invalid; is electromagnetic radiation made up of waves or of particles (photons); and, above all, does the quantum discontinuity represent a necessary boundary to our direct experience of nature?

The revolutionary character of Bohr's proposal can be appreciated if we try to imagine our books arranged neatly in rows on the wall *without any shelves to support them*. Nothing holds Bohr's electrons in their discrete orbitals, except his prescription that they can only occupy orbits which satisfy certain conditions related to an indivisible quantum of action and are not allowed to be anywhere else. This quantum of action, however, Bohr regards as a universal and elementary constant of nature, which itself represents a limit to the usual continuous and causal mechanical account of physical reality. Bohr thus regarded his quantum postulate (the introduction of an abiding discontinuity into the classical picture of physical reality) as reflecting a prescription on the part of nature and substantiating a challenge to our usual awareness of continuities in our description of the external world.

2.2. *Continuity and Discontinuity*

Classical physics, resting on Newton's laws of motion, Maxwell's electromagnetic theory, and statistical thermodynamics, offers an account of physical reality which assumes a continuity of fields of force and motions. Inherent also in classical physics is an assumption about causal interactions in space and time between bodies which are regarded as independent objects. The mathematical formulae used to describe the state of a physical system successfully constituted a theoretical model, in which the terms in the theory corresponded to the elements in the physical system at any time, provided that the system was not interfered with. The movements of the planets, for example, can be traced both in the past and the future. Observation confirms the success of such predictions, but need not alter the state of the system.

The physical system could thus be thought of as entirely independent of observation and theory. Any disturbances introduced by observation and measurement would be, as Bohr would say 'controllable'. If discontinuities are uncovered in the fine structure of matter, however, then this causal, mechanical model is exposed as limited in applicability. Such a challenge occurred at the end of the last century in the consideration of observations of the properties of light-radiation.

All bodies, when heated, emit energy in the form of electromagnetic radiation. Also, the emitting power of a surface is directly related to its absorbing power. A black body is thus not only a very good absorber of radiation, but also a near-perfect emitter. An approximation to an ideal black body can be obtained by drilling a small hole in a closed cylinder. When this body is heated, the radiation from the small hole is equivalent to that from a perfect absorber of light.

Experiments on such cavity radiation produced several interesting results. When the body is heated, the radiation covers a range of energies: for very high and very low wavelengths, the density of radiation is vanishingly small; in between, however, there is a well-defined maximum density. As the temperature is increased, this peak in the curve of energy distributions shifts towards higher energy (shorter wavelength) radiation. Theoreticians initially attempted to explain these observations in terms of classical statistical thermodynamics. Though they were able to match parts of the observed distribution of energies, they were unable to provide a complete picture.

At the end of the nineteenth century, Max Planck, already a distinguished scientist, made a novel suggestion in order to resolve the difficulties in which physicists found themselves. Aware that this proposal would run counter to the precepts of classical physics, if it was indeed an ultimate fact of nature, Planck proposed a 'purely formal assumption': the radiated energy must somehow be treated of not as continuous but as composed of discrete bundles of energy or *Energieelemente*. Planck's intentions are rather difficult to determine. It would seem that he certainly wished to preserve the continuity of radiation, and also of the motions of the oscillating resonators which produce the radiation. His hypothesis was little more than a mathematical ploy to fit theoretical predictions to observed results. All the same, Planck suggested that the size of each *Energieelement* could be related to the frequency of vibration (ν) of the oscillating particles emanating the radiated energy. He thus proposed an equality in terms of a constant which he designated as h. Thus, if the energy of each *Energieelement* was E, then, in Planck's epoch-making equation,

$$\mathrm{E} = h\nu .$$

In a companion paper of 1901 Planck introduced the term *Elementar-quanta* instead of *Energieelement*. The era of quantum physics thus received its emblem. Planck himself called *h* the 'quantum of action', since it had the physical dimensions of action, namely, work multiplied by time. More commonly, however, it became known as 'Planck's constant', though Bohr himself continued to refer to it as the quantum of action.

In his study of black-body theory and the quantum discontinuity, Thomas Kuhn argues that Planck's views in 1901 were, despite his ploys, strictly classical, and that he did not wish to posit a discontinuity as such in radiation.[2] Be that as it may, the vocabulary had been established and a constant had been introduced which related energy to wavelength.

It was Einstein, then an unknown figure, who transplanted the notion of the quantum from the nursery of thermodynamics into the complex world of atomic physics. Einstein's attention was focused on another curious experiment which produced results in conflict with classical prediction, the so-called 'photoelectric effect'. The dislodging of electrons from a surface as the result of electromagnetic radiation had been observed and studied by Heinrich Hertz late in the nineteenth century. Classical theory predicted that the *velocity* of the dislodged electrons should increase in proportion to an increase in the *intensity* of the incident radiation. In fact, however, the velocities remained constant and it was the *number* of electrons being dislodged that was observed to increase. Moreover, one would expect from classical theory that as the energy of incident radiation was reduced, then the dislodging of electrons would simultaneously taper off. Curiously, however, below a certain threshold of energy, no matter how long one applied radiation, no electrons were dislodged at all. There was a startlingly abrupt cut-off point. And, as one varied the frequency of incident radiation above this threshold, so the velocity of the dislodged electrons also increased.

In order to provide a satisfactory theory for these observations, Einstein proposed what he called 'a heuristic viewpoint': what if the incident radiation could be considered as small packets of energy, with the quantity of the energy being directly proportional to the frequency of the radiation? He was thus taking Planck's 'purely formal assumption'

[2] T. S. Kuhn, *Black-Body Theory and the Quantum Discontinuity 1894-1912* (Oxford, Clarendon, 1978), pp. viii, 125-30. Kuhn, departing from Martin Klein's earlier interpretation of Planck's work, provides a much more detailed account of what is noted summarily here.

a step further, shifting the focus of attention from the oscillating particle which produced the radiation to the radiation itself. Einstein was then able to develop a simple mathematical formalism which, at least in theory, accounted for the anomalies of the photoelectric effect.

More important, however, were Einstein's immediate reflections. In a companion paper, published in 1906, he exposed appeal to the quantum as fundamentally counter to the ethos of classical physics: 'the theoretical bases on which Planck's radiation theory rests are different from those of Maxwell's theory'.[3] Planck had not initially intended to quantify light-radiation itself, but Einstein demonstrated that his own 'light-quantum hypothesis' was implicit in Planck's earlier work. In viewing radiation not as a continuous wave, but as composed of small packets of energy (later called photons), Einstein was again shaking the foundations of classical physics.

In 1909 a more celebrated Einstein made the conflict between classical and quantum approaches the theme of his address to a congress of physicists. Many among the scientific establishment could not envisage the abandonment of Maxwell's electromagnetic theory, but Einstein gave judicious support to the notion of the quantum. He was optimistic that more comprehensive theories could be developed and that the light-quanta and light-wave ideas might be reconciled, thus saving the far-reaching claims of classical physics. Like most of his contemporaries at the time, he was dissatisfied with the current situation, which seemed to demand a 'both-and' or an 'either-or' view of wave and particle models for radiation. But whereas Bohr eventually moved towards accepting wave-particle duality, the quest for 'unification' was to preoccupy Einstein for the remainder of his life. It certainly coloured his debates with Bohr about the interpretation of quantum mechanics. In Einstein's later years, therefore, there was a touch of Laplace's arrogance in his admittedly more humble view of physics: nature ought not to exhibit conflicting characteristics like waves and particles, and nor should it display discontinuities of the kind which might disrupt a classical, causal account of physical reality. Once continuity was lost, after all, so also was strict causality: there would be interstices in nature to which the physicist would not be admitted. The notion that theory can offer an independent description of the state of the physical system would be untenable, and mechanical models of a continuously moving system could no longer uncritically be maintained.

[3] A. Einstein, 'Zur Theorie der Lichterzeugung und Lichtabsorption', *Annalen der Physik* 20 (1906), 199; see Kuhn, *Black-Body Theory*, pp. 182-7.

Bohr, who was at first reluctant to accept Einstein's photon-theory of light, was to become the leading advocate of a rather different view of physics. Equally preoccupied with the quest for a more general and unifying point of view in physics, he arrived at a resolution of the paradoxes by accepting that there were limits to our experience of nature as a set of independent objects, and that the bounds of the applicability of classical theory had been exposed.

Before we turn to this, however, we must first consider the physics of Bohr's original quantum theory of the atom in a little more detail.

2.3. *The Bohr Atom and the 'Old' Quantum Theory*

Together with the contributions to physical theory made by Planck and Einstein, several other factors shaped Bohr's insight into the structure of the atom. In his graduate studies, beginning in 1909, he had investigated how new theories of the electron could be related to the observed properties of metals. Bohr could only conclude that the current theories were inadequate: Lorentz's, for example, was that metals consisted of atoms and free electrons. Bohr saw, first of all, that a more general point of view was required, and that it was also necessary to give an account of the way in which electrons behaved *within* atoms.

Other discoveries of the day were potentially related to the structure of the atom: radioactivity, the periodic properties of elements, the regularities exhibited by atomic spectra, and, last of all, the formulae contrived by Rydberg and Balmer to fit observed spectral patterns.

An important set of experimental results was reported by Rutherford in 1911. Upon firing alpha particles (small, positively charged particles) at a thin foil, Rutherford observed a variety of deflections and unobstructed passages. These indicated that the atom was made up of a relatively heavy positive nucleus around which the small negative electrons spun. Just as books without shelves fall to the floor and give off crashing sound-energy, so also the charged and whirling electrons of the Rutherford atom should spiral into the nucleus and emit a flash of ultra-violet light. Rutherford thus suggested that these electrons occupied stable orbitals, in which electromagnetic attraction was balanced by a centrifugal force due to angular momentum. But how was it that they were locked into stable orbits? And why did the orbits exhibit regularity rather than randomness?

In 1912 Bohr moved from Thomson's laboratory in Cambridge to work with Rutherford in Manchester. Rutherford was more of an experimentalist than a theoretician, but with Bohr he made an exception. When he was once asked from which part of the atom a certain type of radiation emanated, Rutherford immediately answered: 'Ask Bohr!' And when pressed to explain how he could hold Bohr in esteem when the latter was primarily a theoretician, he replied: 'Bohr's different. He's a football player!'

While with Rutherford, Bohr attempted to establish the stability of electrons in the atom. He found that classical theory was of no help, whether applied to Thomson's model or to Rutherford's. (Around 1910 A. E. Haas had in fact attempted to give some stability to Thomson's model of the atom by relating frequencies to a ratio of energy and Planck's constant h.) Drawing on the work of Planck and Einstein, Bohr explored the hypothesis that the kinetic energy of rotating electrons was related by some constant to the frequency of rotation. By June 1912 the Bohr atom was beginning to take shape, and Niels wrote to his brother Harald that he had discovered 'perhaps a little bit of the reality'.

On a Christmas card to Harald that year, the recently married Niels and Margrethe wrote: 'one of us would like to say that Nicholson's theory is not incompatible with his own'.[4] Bohr had been considering two papers by John Nicholson on atoms and spectral lines and the patterns observed in stellar spectra. Nicholson was able to explain the observed spectra by making a number of assumptions. One of the fruitful suggestions was that the angular momentum of rotating atomic electrons should be quantized in units of $h/2\pi$. Unlike Nicholson, though, Bohr explored the idea that the emission of radiation was not due to the frequency of rotation of electrons around the nucleus, but rather to a change from a higher energy orbit to a lower energy orbit.

Finally, early in 1913, Balmer's formula came to Bohr's attention. Decades earlier, Balmer had fitted figures to experimental facts by using some ingenious mathematics and an appropriate arbitrary constant. Bohr was able to relate this formula to constants in nature, the mass and charge of the electron, Planck's quantum of action, and transitions between orbit states. Suddenly everything fell into place.

[4] This and the previous quotation are from cards dated 19 June 1912 and 23 Dec. 1912, quoted in *NBCW* 1, pp. 559 and 563. For more detailed accounts of Bohr's development of theories of the atom, see *NBCW* 1; L. Rosenfeld and E. Rüdinger, 'The Decisive Years', in S. Rozental (ed.), *Niels Bohr: His Life and Work* (Amsterdam, North-Holland, 1967), pp. 38-73; and J. L. Heilbron, 'Bohr's First Theories of the Atom', in A. P. French and P. J. Kennedy (eds.), *Niels Bohr: A Centenary Volume* (Cambridge, Harvard University Press, 1985).

In 1913 the twenty-seven-year-old Bohr presented three remarkable papers 'On the Constitution of Atoms and Molecules'. Applying Planck's 'purely formal assumption' and Einstein's 'heuristic viewpoint', he first proposed that the electrons of an atom can only revolve in orbits specified by Planck's constant. This is Bohr's original quantum postulate or 'quantum condition': 'In any molecular system consisting of positive nuclei and electrons in which the nuclei are at rest relative to each other and the electrons move in circular orbits, the angular momentum of every electron round the centre of its orbit will in the permanent state of the system be equal to $h/2\pi$, where h is Planck's constant.'[5] The original quantum condition confined the electrons to particular orbits and hence to specific energy levels. Changes of energy, brought about by transitions between states, show a specific frequency, ν, given by the relation

$$E' - E'' = h\nu,$$

where h is Planck's constant and E' and E'' are the values of energy in the two states concerned. Bohr was legislating against any electron loitering, if not trespassing, in the spaces between such levels. These permitted orbits were given serial numbers, and thus physics acquired the terminology of 'quantum number'.

Though Bohr's original quantum condition was to undergo much revision, the core remained unchanged: an element of limiting discontinuity (or 'individuality' as Bohr inventively put it) had been located at the heart of nature. The new *atoma* (or uncuttable) was no longer the indivisible atom but the quantum of action.

In this same series of papers, Bohr introduced a distinction between the areas in which classical physics remained valid and those in which it could no longer be applied. Referring to the observations of well-defined spectral lines, which were attributed to the movement of electrons between stationary states, Bohr suggested 'That the dynamical equilibrium of the systems in the stationary states can be discussed by help of the ordinary [i.e. classical] mechanics, while the passing of the systems between different stationary states cannot be treated on that basis.'[6] This rule of thumb for comparing calculations was Bohr's version of the consoling adage: 'What you lose on the swings you gain

[5] N. Bohr, 'On the Constitution of Atoms and Molecules', *Philosophical Magazine* 26 (1913), 24 f., *NBCW* 2, pp. 184 f. These papers are reprinted in L. Rosenfeld (ed.), *On the Constitution of Atoms and Molecules* (New York, Benjamin, 1963), and *NBCW* 2, pp. 159-233.

[6] Bohr, 'On the Constitution of Atoms and Molecules', p. 7; *NBCW* 2, p. 167.

on the roundabouts.' That is, what cannot be achieved with classical theory is subjected to quantum considerations, and vice versa. Here we see Bohr's freedom from the hallowed status of classical mechanics and the origin of one version of Bohr's Correspondence Principle, which asserts that classical physics and quantum physics produce equivalent results for high quantum numbers (or on a macroscopic scale). Referring to his work in 1913, Bohr wrote seven years later that 'The first germ of this correspondence principle may be found in the first essay in the deduction of the expression for the constant of the hydrogen spectrum in terms of Planck's constant and of the quantities which in Rutherford's atomic model are necessary for the description of the hydrogen atom.'[7]

Foreshadowed also here is the possibility of using *complementary* accounts for a complete description of the processes being observed. Bohr was aware that the physicist was caught in a strange situation, moving back and forth between two fundamentally different views of nature, on the one hand asserting discontinuity and on the other appealing to continuity. The results of one process were being used to check and balance the results of the other. In 1927 Bohr was to provide a broader account of complementarity in terms of the conditions for unambiguous communication, but here we see the beginnings of the notion and a clue to the underlying consistency in Bohr's vision.

Bohr's approach was astonishingly successful in explaining the immediate problems faced by contemporary physicists. He offered a superbly simple account of the stability of the Rutherford atom, of the pattern of the hydrogen spectrum, and of the significance of the Rydberg constant. But, needless to say, his simultaneous acceptance of two mutually incompatible views of physics left many unhappy with the quantum theory of the atom. Einstein was quite appalled, declaring to Max Born: 'The quantum theory gives me a feeling very much like yours. One really ought to be ashamed of its success, because it has been obtained with the Jesuit maxim: "Let not thy left hand know what thy right hand doeth".'[8]

Bohr was also far from pleased with what he had done. In August 1918 he confessed to O. W. Richardson: 'all the time I am gradually

[7] N. Bohr, Preface to *The Theory of Spectra and Atomic Constitution* (Cambridge, Cambridge University Press, 1922), p. v (henceforth referred to as *TSAC*); see also *NBCW* 4, p. 259. On the evolution of the Correspondence Principle, see section 2.8. below, and *NBCW* 3.

[8] Letter of 4 June 1919, in M. Born (ed.), *The Born-Einstein Letters*, English trans. by I. Born (London, Macmillan, 1971), 10.

changing my views about this terrible riddle which the quantum theory is'.[9] New problems had arisen which made the theory appear at best provisional, and while Bohr continued to affirm his postulates of 1913, he was also ready to acknowledge weaknesses. In 1921 he observed: 'the incomplete character of the theory can be recognized in two obvious ways—not only in the working out of individual details, but also in connection with the grounding of a general point of view'.[10]

The 'general point of view' which Bohr now set about exploring was eventually grounded in epistemological considerations of a transcendental character. Heisenberg and others were also about to formulate the 'new' quantum theory of 1925, which not only provided a general theoretical basis for the physics, but also accounted for the 'individual details' of concern to Bohr. In the meantime, though, Bohr continued to wrestle with the problem of wave-particle duality.

Many studies of Bohr's interpretation of quantum mechanics focus on his notions of complementarity and correspondence. These are indeed the dominant surface features of his vision, as well as the first interpretative strategies to emerge in his published writings, and this chronological survey of quantum physics will help to account for the genesis of the idea of complementarity. The following chapter, however, will present a different perspective on Bohr's vision. Instead of concentrating on the physics which influenced Bohr's thought, we will unearth the more philosophical reflections, from his earliest writings, which underpin his appeal to complementarity. But for the moment let us relish the heady days of physics in the 1920s.

2.4. *The Wave and the Particle*

In his authoritative study of the evolution of Bohr's notion of complementarity, Klaus Stolzenburg concludes that 'Bohr's concept of complementarity developed out of the idea of wave-particle duality'.[11]

[9] Letter to O. W. Richardson, 15 Aug. 1918, Bohr Scientific Correspondence, microfilm 6 (henceforth *BSC* : 6), reprinted in *NBCW* 3, p. 14.

[10] N. Bohr, 'Der Bau der Atome und die physikalischen und chemischen Eigenschaften der Elemente', *Zeitschrift für Physik* 9 (1922), 67, my translation. A slightly different version is given in *TSAC*, translated by A. D. Udden, the original paper having been read in Danish in Copenhagen in mid-Oct. 1921.

[11] K. Stolzenburg, *Die Entwicklung des Bohrschen Komplementaritätsgedankens in den Jahren 1924 bis 1929* (Stuttgart, Ph. D. thesis, 1977), p. 148. For other studies, see K. M. Meyer-Abich, *Korrespondenz, Individualität und Komplementarität* (Wiesbaden, Steiner, 1965); G. J. Holton, 'The Roots of Complementarity', *Daedalus* 99b (1970), 1015-55; D. R. Murdoch, 'Complementarity: A Study of Bohr's Philosophy of Quantum Physics' D.Phil. thesis (Oxford, 1981); and Folse, *The Philosophy of Niels Bohr*.

Three stages were important during this process. First, as noted above, there was conflict between the accepted classical wave-theory of light and Einstein's light-quanta hypothesis. Next, new observations made during Compton's scattering experiments gave empirical evidence for the light-quanta theory. Finally, with the coming of the 'new' quantum theory and Heisenberg's formulation of the indeterminacy relations, a context was established for the discussion of wave-particle duality. We shall examine Bohr's response to each of these developments in turn.

Bohr was extremely reluctant to abandon the wave-theory of light, for it offered a realistic account of the nature of radiation. And, as Chapter 1 intimated, what was real was important to Bohr. Hence it comes as no surprise that up until 1923 Bohr regarded Einstein's notion of light-quanta merely as a useful hypothesis which, on the one hand, resolved the problem of conservation of energy raised in the photoelectric effect, but, on the other, clashed with the evidence of wave-like behaviour from interference effects. Bohr stated: 'the hypothesis under discussion can in no wise be regarded as a satisfactory solution . . . The hypothesis of light-quanta, therefore, is not suitable for giving a picture of the processes, in which the whole of the phenomena can be arranged, which are considered in the application of the quantum theory.'[12] This might be taken to indicate that far from lagging behind Einstein, Bohr had already gone beyond him. Bohr was perhaps not rejecting the light-quanta hypothesis outright, but he did think that it was inadequate on its own. We can assume, therefore, that Bohr was referring to the shortcomings of the light-quanta theory without the complementary wave-theory, and hence shifting debate to the proper use of descriptive concepts. Indeed, in the same section of this paper he also refers to the problem of describing atomic processes in concepts other than those borrowed from classical electrodynamics.

In 1922 A. H. Compton completed experiments which demonstrated the particle character of light. When atoms were irradiated with X-rays it was found that the radiation beam was scattered with alterations in frequency and direction. These changes could easily be accounted for if the process was viewed as an elastic collision between photons and electrons. Indeed, cloud-chamber photographs of the collisions showed effects similar to those we ordinarily observe on a billiard-table. Having

[12] N. Bohr, 'On the Application of the Quantum Theory to Atomic Structure', *Proceedings of the Cambridge Philosophical Society* (*Supplement*) (1924), 35, and *NBCW* 3, p. 492. See also Heisenberg's comments on Bohr's early attitude to light-quanta in the interview with T. S. Kuhn, 15 Feb. 1963, *AHQP* : 49a, p. 15 of transcript.

heard of Compton's paper, Arnold Sommerfeld reported to Bohr in 1923 that 'according to it the wave theory for X-rays is to be definitely given up. I am not yet quite sure whether he is right.' Sommerfeld continued, 'I only wish to call your attention to the fact that here we may possibly expect some very fundamental new information.'[13] For the moment, in fact, Bohr chose to stay with the better established results of diffraction experiments and their implication that light was wave-like. (The wave-nature of particles was not to be demonstrated experimentally until five years later, when C. J. Davisson and L. H. Germer observed the diffraction of electrons by a crystal of nickel.)

Bohr's first attempt to resolve the contradictions between wave and particle theories of light is as instructive now as it was ill-fated then. Along with H. A. Kramers and J. C. Slater, Bohr produced a substantial paper on 'The Quantum Theory of Radiation'. Here the conflict between continuous and discontinuous models of light was resolved by the imposition of a separation between theory and reality. Sticking chiefly to the wave-description, Bohr also employed the light-quanta theory merely as a 'formal' device. Heisenberg recalls the manœuvring thus:

> In 1924 Bohr, Kramers and Slater asserted, first of all, that the wave propagation of light on the one hand, and its absorption and emission in quanta on the other, are experimental facts, which must be made the basis of any attempt at clarification, and not explained away; the fundamental consequence of this state of affairs must, therefore, be taken seriously. They therefore introduced the hypothesis that the waves are of the nature of probability waves: that they represent not a reality in the classical sense, but rather the 'possibility' of such a reality.[14]

This testimony, from one who had first-hand experience of working with Bohr in Copenhagen, helps to clarify the notions of 'virtual' radiation fields and 'virtual' harmonic oscillators, notions which were central to the Bohr–Kramers–Slater paper. Bohr acknowledged his respect for Maxwell's classical theory of radiation, but still maintained that 'it does not seem possible to avoid the formal character of the quantum theory'. Yet he also continued to hold that 'the theory of light-quanta can obviously not be considered as a satisfactory solution of the

[13] Sommerfeld to Bohr, 21 Jan. 1923 (*BSC* : 16), quoted in *NBCW* 5, p. 504.

[14] W. Heisenberg, 'The Development of the Interpretation of the Quantum Theory', in W. Pauli (ed.), *Niels Bohr and the Development of Physics* (London, Pergamon, 1955), 12.

problem of light propagation'.[15] He thus regarded the light-quanta theory as no more than a 'formal' consideration.

By introducing the qualification 'virtual', Bohr was able to step back from a precise description to an 'as if' account. Such 'as if' entities as virtual oscillators were not required to obey the strict classical laws of conservation of energy and momentum, and Bohr was able to reap all the benefits of classical theory without having to accept any of its liabilities. Instead of opting for the solid mechanical realism of Newton and Maxwell, Bohr retreated to the safety of what amounted to a probabilistic and non-causal account of radiation, which offered, by way of consolation, a resolution of the wave-particle dilemma. The problem was now shifted to the theory and away from the reality. One was being asked to suspend belief in the universal applicability of the theory.

Such a procedure is normal enough in theoretical physics. It is arguable that Planck and Einstein proceeded in the same way in their papers employing the quantum of action. But whereas Planck and Einstein saw their work as an intermediate step on the way to a more complete explanation, Bohr was to come to regard this paradoxical state of affairs as irreducible.

What is of particular relevance to the analysis of Bohr's thought, though, is the interest he displayed here in the limits to the applicability of the ordinary spatio-temporal and causal framework which is so commonly used for the ordering of everyday experience. In a supplementary paper Bohr remarks: 'the problem is to what extent the space-time pictures, by means of which the description of natural phenomena has hitherto been attempted, are applicable to atomic processes'.[16] Here we have an indication of Bohr's epistemological interest in the role of conceptual frameworks, a matter to be taken up in Chapter 3. In the classical mechanical picture several elements were tied together: the space-time framework, causality, the conservation of momentum and energy, continuity, and the strong separation between independent physical systems and descriptive concepts. If one element in the scheme became suspect, then so did the entire scheme.

Einstein, it need hardly be said, was troubled by Bohr's subversive remarks about the scope of physics. Bohr seemed to be suggesting that it would be impossible to go beyond probabilist descriptions of atomic

[15] N. Bohr, H. A. Kramers, and J. C. Slater, 'The Quantum Theory of Radiation', *Philosophical Magazine* 47 (1924), 785 ff., and *NBCW* 5, pp. 101 ff.

[16] N. Bohr, 'Über die Wirkung von Atomen bei Stössen', *Zeitschrift für Physik* 34 (1925), 154, and *NBCW* 5, pp. 190, 204. These lines were written after the refutation of the BKS paper.

events. The simultaneous application of causal and space-time frameworks seemed unjustifiable, and as a result energy conservation might have to be abandoned as a fundamental principle. Bohr did not abandon either principle entirely, but he was beginning to question their universal simultaneous application. This only served to add to Einstein's dislike of the original quantum theory of the atom. In a letter to the Borns in 1925, he declared that he feared his attempts to give tangible form to the quanta might be undone. Einstein continued: 'I find the idea quite intolerable that an electron should choose *of its own free will*, not only its moment to jump off, but also its direction. In that case I would rather be a cobbler, or even an employee in a gaming house, than a physicist.'[17] His anxieties were, for the moment, soon allayed: in April 1925, W. Bothe and H. Geiger produced experimental results which presented the theory of virtual oscillators with 'very great difficulties'. The spectre of the Bohr-Kramers-Slater paper was quickly and quietly banished. A gap remained, however, between Einstein's conception of physics and Bohr's scruple about the possible limitations to our ability to order experience. Of more immediate concern also, was the fact that the dualism of wave and particle remained unresolved.

An original solution to this dilemma had been proposed by Louis de Broglie in 1923. He correlated mass (a property of particles) and frequency (a property of waves) by joining Einstein's relativity equation, which relates the mass of a body to energy,

$$E = mc^2$$

and Planck's equation linking energy and frequency,

$$E = h\nu.$$

Combining these two equations, de Broglie suggested that, in theory, a light-quantum could be described by a characteristic frequency proportional to its rest mass. Leaving relativity factors aside, the equation thus becomes, roughly,

$$h\nu = mc^2.$$

He envisaged this rest mass as very small, about 10^{-50}g, so that the velocity of the photon would not differ detectably from the measured velocity of light. Furthermore, he was also able to derive Bohr's original quantum condition from such considerations. While Einstein, through appeal to Planck's equation, had earlier implied that light-frequency could be related to particle-like momentum, de Broglie was now

[17] Letter of 29 Apr. 1924, *The Born-Einstein Letters*, p. 82.

proposing a wave-particle equivalence not only for light, but for electrons and matter in general. He thus claimed to have saved both 'the corpuscular and undulatory characters of light'.[18]

For all that it reconciled the dualism of wave and particle, de Broglie's suggestions also stated the paradox more starkly: wave and particle are more different than the proverbial chalk and cheese. New ideas were still required. Kramers, perhaps reflecting a growing consensus, had mooted the procedural principle that quantum-physicists should restrict their attention to 'only such quantities as allow of direct physical interpretation'.[19] This idea also appeared in the Bohr-Kramers-Slater paper, as well as occupying a central position in Heisenberg's fashioning of the new quantum theory. Kramers's suggestion had in fact been anticipated by Bohr in 1913:

> The subject of direct observation is the distribution of radiant energy over oscillations of the various wavelengths. Even though we may assume that this energy comes from systems of oscillating particles, we know little or nothing about these systems. No one has ever seen a Planck's resonator, nor indeed even measured its frequency of oscillation; we can observe only the period of oscillation of the radiation which is emitted.[20]

For an anti-realist, of course, wave-particle duality is not a problem. That it was of concern to Bohr, therefore, is an indication of his realist philosophy of science. Even at this early stage of his thinking, he was ready to make a distinction between atomic objects and the effects which such objects have on interaction with observing apparatus. If one recognizes that it is only possible to talk about what can be observed, this is not, in Bohr's case, to deny the reality of the objects which are part of the system-apparatus interaction.

In 1925, Heisenberg, who declares himself to have been influenced by Einstein rather than by Bohr or Kramers in his youthful positivism, embarked on an attempt 'to establish foundations for a quantum-theoretical mechanics which is exclusively based on the relationships between magnitudes which are, in principle, observable'.[21] At the same

[18] L. de Broglie, 'A Tentative Theory of Light Quanta', *Philosophical Magazine* 47 (1924), 446-58.

[19] H. A. Kramers, 'The Quantum Theory of Dispersion', *Nature* 114 (1924), 311.

[20] N. Bohr, 'On the Spectrum of Hydrogen', in *TSAC*, pp. 10 f., and *NBCW* 2, pp. 292 f.

[21] W. Heisenberg, 'Über quantentheoretische Umdeutung kinematischer und mechanischer Beziehungen', *Zeitschrift für Physik* 33 (1925), 879; see also p. 880. For a detailed account, see E. MacKinnon, 'Heisenberg, Models and the Rise of Matrix Mechanics', *Historical Studies in the Physical Sciences* 8 (1979), 137-85.

time, but working independently, Erwin Schrödinger followed through the implications of the de Broglie-Einstein wave-theory of moving particles. These twin formulations of the new quantum theory thus entailed a reconsideration of the status of wave and particle notions as well as the problem of the relationship between theoretical description and observed reality. Were wave and particle simply descriptive terms? Had physics found itself distanced from reality? Bohr struggled with the same old questions, but now in a new climate of thought.

2.5. *The New Quantum Theory*

Heisenberg had been one of Sommerfeld's students in Munich. In late 1923 he became an assistant to Born in Göttingen. At this time Born was working on problems related to the frequencies exhibited in atomic spectra, and it was becoming clear to him that a new kind of mechanics, which he first termed 'quantum mechanics', was required. The problem was that the origins of atomic spectra now appeared to be the result of a transition between two permitted electronic orbits or stationary states. Classical mechanics could only describe motions in one orbit or another; it could not offer a theory which included transitions between states. Classical kinematics followed the path of a particle through time, but such a path was impossible to observe in the case of an electron, and so Heisenberg shifted his attention to what could in principle be observed, namely, differences in energy levels.

Investigations of the spectra of particular elements produced light-emission patterns which provided information about the frequency and intensity of the light-radiation: frequencies indicated the energies required to move an electron between quantum levels, while respective intensities reflected the populations of electrons moving between particular levels. This was the only data Heisenberg allowed himself to work with, for it was based on observable magnitudes in accordance with his principle for the foundations of quantum mechanics. Heisenberg found himself faced with endless permutations of sets of pairs of frequencies and intensities relating to transitions between particular stationary states. Upon expressing these quantities as a mathematical series, he then had to find a rule which both connected observed results and which might predict the results of future observations.

Heisenberg moved from considerations of a virtual oscillator to the equivalent of a two-dimensional table of information obtained from, and obtainable from, observations. In manipulating these tables, he was forced to impose a new (to him) and disturbing multiplication rule: in some circumstances, for any paired components of his tables (say p and q), the result of multiplying p by q had to be made different from that of multiplying q by p. Not only that, but if the theory was to fit the facts, the difference between these two calculations had to be made a factor of Planck's constant, h. Thus Heisenberg 'invented' (his own word) the new quantum theory.[22] This achievement was so extraordinary, even by the standards of the time, that Born described it to Einstein as 'sehr mystisch'.[23]

As Born and Jordan were quick to notice, Heisenberg had accidentally stumbled upon the principles of matrix algebra. The multiplication rule which he had hammered together in his Procrustean manner was well known to those familiar with the theory of matrices. This rule describes the 'non-commuting' character of matrices and their elements. That is, the ordered pair p and q is said not to commute when

$$pq - qp \neq 0.$$

Working with Heisenberg at Göttingen, Born and Jordan soon revised the matrix quantum theory and produced a 'sharpened quantum condition' superior to that introduced by Bohr in 1913. Where Bohr had quantized *states* (electron levels) in terms of Planck's constant, the new theory also quantized the *processes* between states in terms of Planck's constant and the complex number i ($\sqrt{-1}$).

Within the space of a year, two further formulations of a new quantum theory were produced by Schrödinger and P. A. M. Dirac. Where Heisenberg, under Bohr's influence, had worked with observables like intensity and frequency, Schrödinger explored de Broglie's notion of the wave-nature of particles. This led to the production of the 'Schrödinger equation' and the beginnings of wave-mechanics. The heart of this differential equation is a wave-function, but it also implicitly included Planck's constant and i. Initially, Schrödinger interpreted this to indicate that 'material points consist of, or are nothing but, wave-systems'.[24] The solutions of this equation, when applied to the problems of electron orbitals and transitions, were as

[22] W. Heisenberg, *The Physical Principles of the Quantum Theory* (Chicago, Chicago University Press, 1930), 11.

[23] Letter of 15 July 1925. 'Sehr mystisch' is weakly rendered as 'rather mystifying' in *The Born-Einstein Letters*, p. 84.

[24] E. Schrödinger, 'An Undulatory Theory of the Mechanics of Atoms and Molecules', *Physical Review* 28 (1926), 1049.

satisfactory as the solutions of Heisenberg's matrix-mechanics. In a later paper of 1926, Schrödinger was able to demonstrate the mathematical equivalence of the wave and matrix formalisms in the new quantum mechanics. Dirac, employing a sophisticated mathematics proposed by Poisson, produced a formalism similar to Heisenberg's but with a more general non-commuting effect.

The new quantum mechanics proved immediately successful in a range of applications, but fundamental questions about the new theory remained unresolved. First, a completely consistent mathematical framework for the new theory had yet to be fashioned. Secondly, it was hard to swallow the fact that both Heisenberg and Schrödinger had virtually plucked their theories out of the air, often counter-intuitively and against the best traditions of current physics. Thirdly, what physical basis did theoretical terms have: did Schrödinger's equation imply that physical reality was in fact made up of waves and not particles? While this was a tempting implication, how could one account for the 'unreal' element, *i*, in the complex wave-function? Further, Schrödinger's wave moves in a Hilbert space of 3n dimensions, where 'n' is the number of particles in the system. Such a wave-function, as Bohr, Born, and Heisenberg saw, could not be construed directly to represent familiar physical reality. Finally, and a little later, the question would be asked whether specific measurements of position in atomic experiments entailed some sort of 'reduction of the wave packet' at the moment of interaction between measuring apparatus and sub-atomic process? Could what was spread out and wave-like, containing a whole series of probable results of measurement in other words, suddenly be collapsed into one specific result?

Contrary to those who took the wave-function at face value, Max Born proposed that it could only be related to *our* measurement of a physical system, rather than to the physical system itself: 'The new mechanics doesn't answer, as before, the question "How does a particle move?" but the question, "How probable is it that the particle moves in a particular way?" '[25] This suggestion, in a slightly revised form, was to become the basis of the probabilistic interpretations formulated in Copenhagen.

The 'probabilistic' interpretation regards the sub-atomic processes as not being governed by deterministic laws. A 'statistical' interpretation, on the other hand, retains the ideal of classical physics and considers the

[25] M. Born, 'Das Adiabatenprinzip in der Quantenmechanik', *Zeitschrift für Physik* 40 (1926), 167.

wave-function in the quantum formalism as referring to an 'ensemble' of similar systems, rather than to an individual system. Hence it permits the same degree of control as achieved in classical statistical thermodynamics. This statistical interpretation was the bastion of the post-quantum classicism which Schrödinger and Einstein later sought to defend. (Though Born and Bohr both refer to their interpretation as 'statistical', they give that term a 'probabilist' meaning.)

In Easter 1926 Heisenberg moved to Copenhagen to work with Bohr. The crisis raised by the new quantum mechanics became even more acute the following year when Heisenberg established the indeterminacy relations which became known as his 'uncertainty principle'. Even with ideally perfect instrumentation, the in-built consequences of the quantum formalism meant that there was a limit to the sharpness with which one could simultaneously measure paired observables such as position-momentum or time-energy. This limitation proved to be dependent on the quantum of action, h.

If the quantum theory was complete, then the indeterminacy relations marked the end of classicism in physics, in as much as the classical approach aspired to certitude of causal connections within an absolute space-time framework. In classical astronomy, for example, one could make a series of measurements of planetary motions and positions and, using Newton's laws, determine with certainty (within the limits of contingent experimental error) the past and future behaviour of these heavenly bodies. In other words, as long as the mechanical system is 'closed' (or free of external perturbation) then the *state of the system* can be known at any time through appeal to the parameters which define it. In classical theory, the usual parameters defining the state of a mechanical system are position and momentum. As long as the system is 'closed', and assuming the conservation of energy and momentum, then at any time we can claim to know the objective state of the system. The 'state function' in classical physics was seen to correspond to the state of the system.

While classical physics remained successful on this macroscopic scale, the new quantum theory dampened hopes for a similar achievement at the sub-atomic level because of the effects of discontinuous change on this scale. *Experimentally*, it was impossible simultaneously to get the information one needed to make such predictions. As Heisenberg's thought experiment with a gamma-ray microscope demonstrated, 'the more accurately one determines the position [of a particle], then the more inaccurate is the determination of its momentum, and vice versa'.[26] *Theoretically*, according to Born's probabilistic interpretation, one could no longer follow isolated motion with classical certitude. The state function now describes only the likely state of a system/apparatus interaction; prior or subsequent states of the system itself are unknowable. The state functions can only provide us with the likely results of other experimental interventions. The 'objectivity' of quantum physics, therefore, is concerned with the objectivity of observations rather than with the objectivity of the unperturbed state of the system itself. Indeed, the idea of a quantum system having a kind of undiscovered but determinable state of its own may have to be abandoned. We are unable to picture the object apart from observations which entail interaction with measuring apparatus, and this in turn implies both the acceptance of acausal interactions and the employment of classical descriptive concepts.

Heisenberg at this time saw his indeterminacy relations as bolstering a Machian positivism: 'Physics must be confined to a formal description of the relation between perceptions.'[27] Bohr took issue with Heisenberg's point of view, even if he eventually agreed with the derivation of the indeterminacy relations. For Heisenberg, it was the mathematical clarity that was of prime importance, and the fundamental problem of wave-particle duality was not immediately so relevant. Such a stance, in Bohr's mind, left quantum theory incomplete, for it failed to resolve the incompatibility of the wave and particle pictures.

After relentless argument with Bohr, which left him in despair and tears, Heisenberg added a postscript to his paper, to the effect that Bohr would soon provide a refinement of the implications of his work. Bohr's position, as Jammer has pointed out, is supported by the fact that Heisenberg's derivation of the indeterminacy relations depended on appeal to the Einstein-de Broglie equations which linked wave and particle attributes together and implicitly presupposed wave-particle duality rather than resolving it.[28] Heisenberg's restriction of physics to what was measurable certainly confirmed Bohr's instincts as to what was definable, but Bohr's focus is the more fundamental. The indeterminacy in establishing conjugate quantities is not so much a consequence of what can be measured as it is of what can be defined.

[26] W. Heisenberg, 'Über den anschaulichen Inhalt der quantentheoretischen Kinematik und Mechanik', *Zeitschrift für Physik* 43 (1927), 175.

[27] Ibid., p. 197. Heisenberg later modified this position: see P. Heelan, *Quantum Mechanics and Objectivity* (The Hague, Nijhoff, 1965), 137-55.

[28] M. Jammer, *The Philosophy of Quantum Mechanics* (New York, Wiley, 1974), 69.

Bohr perceived that a situation had now been established in which it was impossible to use both wave and particle descriptions simultaneously, and yet in which both descriptions were required if a complete account of physical reality was to be given. Heisenberg recalls Bohr's argument with him as follows:

> The main point was that Bohr wanted to take this dualism between waves and particles as the central point of the problem. . . . I, in some way, would say, 'Well we have a consistent mathematical scheme and this consistent mathematical scheme tells us everything which can be observed. Nothing is in nature which cannot be described by this scheme' . . . Bohr would not like to say that nature imitates a mathematical scheme . . .
>
> I learned that the thing which I in some way attempted could not be done. That is, one cannot go entirely away from the old words because one has to talk about something. . . . I realized, in the process of these discussions with Bohr, how desperate the situation is. On the one hand we knew that our concepts don't work, and on the other we have nothing except the concepts with which we could talk about what we see.[29]

Heisenberg's work was not the origin of Bohr's principle of complementarity, but it stirred Bohr to give further attention to the problems of interpreting the quantum formalism and the conditions for the proper use of descriptive concepts.

Attempts were made to restore classical clarity by treating wave-particle duality as conceptual rather than ontological and thereby dismissing the paradox of it. Jordan and Klein and Wigner, for example, published papers in which the mathematical formalisms rendered the choice of wave or particle descriptions irrelevant. Bohr, for different reasons, also appealed to a distinction between formalism and fact. But Bohr's instinct was not to distinguish his way *out* of the babel of descriptive concepts. On the contrary, he gave a primary place to our abiding observation-terms in his epistemological lesson on the proper definition of observations.

Heisenberg's discovery constituted a mathematical description of the bounds of sense and experimental observation. Bohr had anticipated him in various remarks on the philosophical implications of the quantum, but now Heisenberg had presented him with a potent warrant. If the difficulties of interpretation had become more acute, their resolution could also become more radical.

[29] W. Heisenberg, interviewed by T. S. Kuhn, 25 and 27 Feb. 1963, *AHQP*, tapes 52a, p. 15, and 52b, p. 26 of transcripts.

2.6. *Bohr's Interpretation*

The indeterminacy relations represented a dramatic limitation to the consistency of classical accounts. And towards the end of 1927, as Bohr voiced his 'complementarity' interpretation of the new quantum theory for the first time, he paid tribute to his assistant: 'An important contribution to the problem of a consistent application of these [quantum-theoretical] methods has been made lately by Heisenberg.'[30] This paper on 'The Quantum Postulate and the Recent Development of Atomic Theory' was delivered at the Volta Centenary in Como, again in Copenhagen, and yet again at the 1927 Solvay Conference for physics. It was Bohr's first major statement of his interpretation of the new quantum theory. The problems he investigated were all related to the breakdown of classical mechanics: the separation of causality and determinism, the limitation of space-time frameworks, the relationship between classical and quantum physics, and the physical significance of the elements in the mathematical formalism of quantum theory.

The rapid refutation of the Bohr-Kramers-Slater paper proved the danger in any strategy which compromised with classical theory. As far as Bohr was concerned, the key assumptions of classical theory, including the simultaneous application of causality and continuity in space and time, now had to be challenged head-on. In 1925 he wrote: 'in the general problem of the quantum theory, one is faced not with a modification of the mechanical and electro-dynamical theories describable in terms of the usual physical concepts, but with an essential failure of the pictures in space and time on which the description of natural phenomena has hitherto been based.'[31] Bohr's correspondence at this time discloses a similar preoccupation with the role of experiential concepts, conceptual frameworks, and the limited applicability of mechanical paradigms. He wrote to Rutherford, his old mentor, that he was planning 'to describe the nature of the outstanding difficulties which offer such uncomforting indications of essential

[30] N. Bohr, 'The Quantum Postulate and the Recent Development of Atomic Theory'. As Stolzenburg has shown, there are several versions of this so-called Como Address, which Bohr delivered at the Volta Centenary Congress in Como on 16 Sept. 1927 (and on other occasions): see Stolzenburg, *Die Entwicklung des Bohrschen Komplementaritätsgedankens*, pp. 147 ff. This quotation is from the version printed in *ATDN*, p. 57.

[31] N. Bohr, 'Atomic Theory and Mechanics', *ATDN*, pp. 34 f. See also pp. 42, 45, 47, 50, for Bohr's further remarks on space-time frameworks. The original Danish paper was revised in the light of Heisenberg's matrix formulation: see Bohr's note in *Nature* 116 (1925), 845, n. 1.

deficiencies in our usual description of natural phenomena'. And to Bohr, in a discussion of coupling interactions, he noted that certain assumptions 'excluded the possibility of a simple description of the physical events in terms of intuitive pictures [anschaulicher Bilder]'.[32]

Bohr had been privately pondering these philosophical problems for some time. In 1922 he wrote a long letter to the Danish philosopher H. Høffding, a family friend, which reveals his instincts about the implications of the quantum for human knowing. Note also Bohr's interest in models and analogies in science, especially in discourse about novel experience and theory:

> The question about the role of analogy in scientific research is clearly an essential feature of any work in natural science, even if it is not always obvious. It is often possible to employ a mathematical or geometrical model which handles the problems in question in such a way that the account acquires almost a purely logical character. In general, however, and particularly in some new fields of investigation, one must remember the obvious or likely inadequacy of pictures: as long as the analogies show through strongly one can be content if their usefulness—or rather fruitfulness—in the area in which they are being used is beyond doubt. Such a state of affairs holds not least from the standpoint of the present atomic theory. Here we find ourselves in the peculiar situation that we have obtained certain information about the structure of the atom which may surely be regarded as just as certain as any one of the facts in natural science. On the other hand, we meet with problems of such a profound kind that they seem to defy solution: it is my personal opinion that these difficulties are of such a nature that they hardly allow us to hope that we shall be able, within the world of the atom, to carry though a description in space and time that corresponds to our ordinary sensory images.[33]

Similar musings are to be found throughout Bohr's early correspondence, and will be referred to below in the account of the transcendental character of his philosophy. For the present, though, our focus will be on *what* Bohr had to say, rather than on *why* he said it.

The philosophical nature of Bohr's early papers on the interpretation of quantum theory remains obvious none the less. Bohr's first written reference to complementarity occurs in some lapidary notes from July 1927 (dated, perhaps in his excitement, as 1926) which begin with what must be his basic premisses: 'All information about atoms expressed in

[32] Letter to Rutherford, 28 Apr. 1925, *BSC* : 15; letter to Born, 1 May 1925, *BSC* : 9; see also the letter to Fowler, 21 Apr. 1925, *BSC* : 10. These are quoted in *NBCW* 5, pp. 489, 85, 310 f., and 81 respectively.

[33] Letter to Høffding, 22 Sept. 1922, *BSC* : 3, from the Danish (following Murdoch's translation). See also Bohr's letter to J. Franck, 29 Sept. 1922, *BSC* : 2.

classical concepts. All classical concepts defined through space-time pictures . . . Complementary aspects of experience that cannot be united into a space-time picture based on the classical theories . . .'[34] Bohr formally presented this notion of complementarity to the world's leading physicists (with the exception of Einstein) gathered in Como in 1927. Although it is not certain that the actual text of Bohr's address has survived, the version which Stolzenburg regards as most likely (or most like) again begins with the issue of classical concepts and theoretical frameworks:

Characteristic of the quantum theory is the acknowledgement of a fundamental limitation in our classical physical ideas when applied to atomic phenomena. Just on account of this situation however we meet with intricate difficulties when attempting to formulate the contents of the quantum theory in terms of concepts borrowed from the classical theories. Still it would appear that the essence of the theory may be expressed through the postulate that any atomic process open to direct observation involves an essential element of discontinuity or rather individuality completely foreign to the classical ideas and symbolized by Planck's quantum of action. This postulate at once implies a resignation as regards the causal space time coordination of atomic phenomena.[35]

In his inventive English, Bohr then immediately turns to the problem of measurement, and thus also the *observation* of wave-particle duality: 'Indeed our usual space time coordination rests entirely on the idea of tools of measurements the interaction of which with the phenomena to be observed may be neglected. . . . According [to] the quantum postulate, however, every observation of atomic phenomena will just involve an individual process, resulting in essential interaction. We cannot therefore speak of independent tools of measurements.' What Bohr means by 'phenomena', 'measurement', and 'individual' will be clarified in section 2.9 below. The point he is making here is that, up until now, physicists have been able to 'control' or compensate for any effects due to interaction between measuring apparatus and the objects being observed. One can no longer claim such a strong distinction

[34] Notes in Danish, dated 10 July 1926 (although clearly from 1927), *MSS* : 11, reprinted and translated in *NBCW* 6, pp. 59 ff.

[35] There are hundreds of pages of notes for this address, from which Stolzenburg has identified several versions and classified them. This and the following quotations are from a manuscript of 8 pages in Bohr's own handwriting, dated 13 Sept. 1927 (3 days before he spoke at Como), entitled 'Fundamental Problems of the Quantum Theory'. See Stolzenburg, *Die Entwicklung des Bohrschen Komplementaritätsgedanken*, pp. 156 ff., and Bohr Scientific Manuscripts, microfilm 11 (henceforth *MSS* : 11), reprinted in *NBCW* 6, pp. 75 ff.

between observer and observed, since the disturbance to what is being observed cannot be controlled. If this is true, Bohr suggests, then the problem of wave-particle duality is posed both in nature and, consequently, in the theoretical frameworks we apply as a result of our experimental observation of nature.

The search for a new theoretical physics was not all that Bohr had in mind in this address at Como. He also explored the connections between nature and theory, between experience and language, between mechanics and 'picture concepts'. These reflections on the necessary conditions for the unambiguous description of experience led Bohr to discover an 'epistemological lesson' to which he would appeal over and over again. Here are the beginnings of his transcendental philosophy.

Bohr thus set about introducing a switch between the rigid spatio-temporal and causal framework of classical physics and a novel 'complementarity' framework. Dissociating himself from Heisenberg's position at the time, Bohr then hints at the importance of an epistemological enquiry into the foundations of physics: 'I shall try however here to present the fundamental problem at issue from a somewhat different point of view, taking the starting point in the analysis of the most elementary features of our experience regarding atomic phenomena.' The reasoning in this address is not explicit. Bohr almost proposes his solution as if by pragmatic fiat. Let there be complementarity! If the duality of wave and particle is a problem, it is because of the rigidity of our frameworks: what we observe as a particle cannot possibly be a wave, and vice versa. If we accept that the mode of observation limits the account of what is observed, then we can tolerate such a duality.

After all, Bohr seems to argue, we cannot simultaneously measure wave and particle properties in the same experiment. Whether we choose a diffraction-experiment or a scattering-experiment, we are observing different aspects of the processes under investigation. And so Bohr resolves the problem of wave-particle duality in one stroke: 'we are not dealing with a choice between a wave or corpuscular theory of light. Indeed the wave and corpuscular ideas are able only to account for complementary sides of the phenomena.'[36] And, as a result of the implications of quantum mechanics, it will be impossible in any measurement to specify precisely and simultaneously positions and momenta, times and energies. Without these specifications, the possibility of a causal account is also lost. Bohr thus continues: 'Indeed,

[36] *NBCW* 6, p. 76.

we may say that according to the quantum theory the possibility of a space time coordination is complementary to the possibility of a causal description.'[37]

It is worth noting that in the Como lecture, as the last quotation indicates, Bohr seemed to make a special connection between 'causality' and descriptions which included the conservation of energy and momentum. Similarly, he made a parallel connection between 'definition' or 'observation' and descriptions which refer to the space-time framework. Note also that Bohr does not want to do away with causality altogether: the very fact of observation rests on some sort of interaction, albeit uncontrollable, between quantum system and the observing apparatus. Such an interaction, however, interrupts any space-time picture of the quantum system. In this sense he speaks of causality and space-time as being complementary. Once their simultaneous applicability is lost, however, the strongly objectivist classical mechanics of the state of the system is also no longer completely tenable. The well-known complementarity of wave and particle descriptions is thus connected with the lesser known complementarity of space-time and causality both at the beginning and at the end of Bohr's reflections.

The obscurity in Bohr's thinking here can be partly removed by considering the notion of 'individuality'. In classical physics one was permitted to make a series of measurements without in any way disturbing the system being measured. Observations did not interfere with motion, and a series of observations built up a picture of movement through space and time, so that a causal account could be given of why a particular motion occurred and would occur. In quantum physics, however, it seems that each experiment describes a particular event in which one aspect or another of an interaction between atomic system and measuring agency can be investigated. Each experiment interferes with the processes under observation to such a degree, moreover, that the possibility of a series of detached observations is out of the question. This is not just a matter of contingent experimental error, but a consequence of the discontinuity in nature symbolized by the discovery of the quantum of action: though we might be able to determine position at one moment, we will be denied detached observation of position at another. This is because the system under observation has been irredeemably interfered with due to the nature of the discontinuity arising from the existence of the quantum of action. Without a second

[37] Ibid., pp. 78 f.

undisturbed observation of the isolated system, we are also denied the possibility of offering a causal account. Indeed, Bohr argued, in quantum physics one can only consider the individual events of interaction between classical apparatus and atomic process. This is the whole reality. Once we interfere in the space-time framework, there is no sequence of events, as it were, to observe and connect causally. Just as Bohr regarded wave and particle as mutually exclusive concepts, so also he suggested that space-time and causality must be regarded as complementary in quantum physics. Just as the combining of wave and particle descriptions gave us a more complete account of what was being observed, so also could we combine space-time and causality.

Twenty years later, reviewing his first statement of these ideas, Bohr summarized his thesis in two carefully worded paragraphs. (Note that in using the word 'phenomena' here, Bohr refers to observations obtained under particular experimental conditions, whereas in his 1927 address he had used it both to refer to the atomic events themselves as well as to our observations of them.) Bohr recalls his early argument thus:

> I advocated a point of view conveniently termed 'complementarity', suited to embrace the characteristic features of individuality of quantum phenomena, and at the same time to clarify the peculiar aspects of the observational problem in this field of experience. For this purpose, it is decisive to recognize that, *however far the phenomena transcend the scope of classical physical explanation, the account of all evidence must be expressed in classical terms* . . .
>
> This crucial point . . . implies the *impossibility of any sharp separation between the behaviour of atomic objects and the interaction with the measuring instruments which serve to define the conditions under which the phenomena appear.*[38]

Bohr shows that the roots of his thinking lie in a consideration of the conditions governing our accounts of experimental evidence. All the same, his 'must' seems rather dogmatic in this context. The following chapter sets out in detail the reasons for his argument relating to the conditions for unambiguous communication.

This argument is not so apparent in the Como address, however. What is clear is that, for Bohr, 'complementarity' connotes a more general framework or point of view in which to fit together previously conflicting aspects of our participation in nature. Secondly, the prescription of complementarity entails a renunciation of causal determinism in the realm of quantum physics and a resolution of

[38] *APHK*, pp. 39 f., Bohr's emphasis.

wave-particle duality. Thirdly, the consideration of the conflict between classical and quantum physics leads to a revised version of the Correspondence Principle, which is expanded to deal not just with comparing calculations by quantum or classical methods, but also with the application of classical concepts to quantum observations. Lastly, Bohr is able to give an alternative account of the nature of measurement in quantum physics. We shall now examine these elements in Bohr's early interpretation.

2.7. *Complementarity of Descriptions*

Bohr's 'point of view' of complementarity was intended to 'provide a frame wide enough to embrace the account of fundamental regularities of nature which cannot be comprehended within a single picture'. It 'offers a logical means of comprehending wider fields of experience' and 'a wider view and a greater power to correlate phenomena which before might even have appeared as contradictory'. Complementarity is not just a relationship between modes of description, however, but also a 'mode of description' in itself. One may thus speak of 'complementary modes of description'.[39]

Furthermore, writes Bohr, the notion of complementarity entails exhaustive and complete description in so far as it embraces mutually exclusive descriptive concepts drawn from theoretical frameworks. It allows us to consider both sides of the coin together (though this analogy breaks down completely: in dealing with quantum objects such as electrons we cannot presume to think of 'sides of a coin' so much as results of observations which depend on the choice of apparatus). What we see owes as much to where we stand as it does to the minting of the coin. This should not be taken to mean that our perception creates the reality, nor that the reality is not there if we are not perceiving it, nor that Bohr is simply making the point that our knowledge-claims must be restricted to our experiences. For us to talk about the properties of light, for example, we must use the concepts of our everyday language, or their refinements: we discover information about these properties by allowing light to interact with macroscopic apparatus. Just as we require more than one experimental arrangement to describe these properties exhaustively, so also we need to accept the imposition of complementarity.

[39] See respectively *Essays*, p. 12, *APHK*, pp. 78, 5 f., *ATDN*, p. 10, and *APHK*, pp. 93, 101.

In classical physics, it might be objected, we also employ series of experiments in order to fill in the details of a particular system under investigation. In classical physics, however, continuity of change of state, given the assumption of one-to-one correspondence between the terms in the state function and the various properties of the state of the system, goes hand in hand with a continuity of descriptions of observations. Observations which are conjoined in classical physics are, because of the discontinuity introduced by the quantum of action, no longer possible in quantum physics. Hence Bohr claims that a totally different situation prevails:

> A most conspicuous characteristic of atomic physics is the novel relationship between phenomena observed under experimental conditions demanding different elementary concepts for their description. Indeed, however contrasting such experiences might appear when attempting to picture a course of atomic processes on classical lines, they have to be considered as complementary in the sense that they represent equally essential knowledge about atomic systems and together exhaust this knowledge.[40]

Bohr's concern here is with the use of concepts, the bounds of ordinary experience, and the individuality of observations on the quantum scale. Whether his views are idealist, positivist, realist, or whatever, will be discussed below. For the present, let us give some of Bohr's clarifications of the notion of complementarity.

At times, especially in one important paper in 1929, Bohr dropped the term 'complementarity' and 'preferred the term "reciprocity" . . . to denote the relation of mutual exclusion characteristic of the quantum theory with regard to the application of various classical concepts and ideas'.[41] Bohr's coat of arms, designed in 1947, offers another clue: it contains the motto, 'Contraria sunt Complementa', and the Chinese Yin-Yang symbol of interlocking opposites forming a perfect circle. The implication is that it is in a circular as well as in a dialectical manner that the complementary descriptions offer an exhaustive account. Bohr does not speak of circularity as such, but he decisively withdraws his notion of complementarity from the courts of classical physics: 'there is never any question of a complete description similar to that of the classical theories'.[42]

[40] *APHK*, p. 74. Bohr uses terms like 'exhaustive description' frequently: see *APHK*, pp. 40, 88, 90, and *Essays*, pp. 4, 12.
[41] *ATDN*, p. 19; see also pp. 18, 67, 94, 95, and *Essays*, p. 43.
[42] *ATDN*, p. 17.

Bohr claimed that complementarity was the new, more general framework which superseded, as well as partly embraced, the narrower notion of causality in order to give coherence to a wider range of experimental observations. Complementarity provided 'a frame wide enough to embrace the account of fundamental regularities of nature'; or, a 'framework sufficiently wide for an exhaustive description of new experience'.[43] These prosaic generalities become more cutting, however, when seen to entail a direct challenge to the ubiquitous application of causality and to the notion that physics can provide deterministic prediction in every area of its concern. The new physics, says Bohr, 'forces us to replace the ideal of causality by a more general viewpoint usually called "complementarity" '.[44]

Put bluntly, at the bounds of experience *complementarity takes over the role of causality* and provides a more general point of view. The sequential framework of cause and effect is embraced by a more circular account. Instead of treating observations in a series, Bohr is suggesting that we now have to go back and forth between sets of observations. Yet causality and space-time continuity are not done away with: they are seen as reciprocal, rather than conjoint, modes of description. In the essay on interpretation of quantum theory which Bohr regarded as his best, he emphasizes this transition: 'Summarizing, it may be stressed that, far from involving any arbitrary renunciation of the ideal of causality, the wider frame of complementarity directly expresses our position as regards the account of fundamental properties of matter presupposed in classical physical description, but outside its scope.'[45] As Bohr had put it in an earlier essay 'On the Notions of Causality and Complementarity', one must regard 'complementarity . . . as a rational generalization of causality'.[46]

In classical physics, causality and space-time were combined to offer a single mechanical description of a physical system, independent of the interactions by which we observed the state of that system. Accepting the quantum postulate, however, requires that we can only describe

[43] *Essays*, p. 12, and *APHK*, pp. 65-8, 88.

[44] N. Bohr, 'Causality and Complementarity', *Philosophy of Science* 4 (1937), 291. See also 'Quantum Physics and Philosophy: Causality and Complementarity', in *Essays*, pp. 1-7. The point about the connection between complementarity and causality is made in E. Scheibe, *The Logical Analysis of Quantum Mechanics* (London, Pergamon, 1973), 14, 30.

[45] *Essays*, p. 7.

[46] N. Bohr, 'On the Notions of Causality and Complementarity', *Dialectica* 2 (1948), 317.

microphysical systems in terms of their interactions with apparatus: the state of the system after such an interaction is not definable, since the resulting change of state due to the interaction proceeds discontinuously. The unhindered path of an electron, for example, cannot be followed in space and time while a causal interaction is employed to observe it. If we choose to use a space-time description, then we must renounce a determinist causal account. But then we are left speaking of what cannot be observed. In this sense causality and space-time are mutually exclusive. They can only be combined, says Bohr, by invocation of complementarity.

The concept of complementarity is an elusive one, but not because it is complex. Quite the opposite, it is difficult to grasp because it is so utterly simple. The main problem is that the old habit of thought, the classical framework linking continuity, causality, and space-time, dies hard. Bohr was proposing a simple alternative. He confessed to his good friend C. G. Darwin (whom he used to call 'the grandson of the real Darwin') when he was giving all his attention to the complementarity interpretation: 'I get afraid that it is all too trivial.' And many years later, writing to Pauli about the various reactions to his notion of complementarity, Bohr remarked: 'To my mind the situation is far more clear than generally assumed, and such tools as three-valued logics I consider rather as complications.'[47]

According to Bohr, 'complementarity' has the same function in organizing our outlooks as 'causality' does, though this is not to say that the two operate in the same way or at the same level. When one speaks of causal connections between action and reaction, rainfall and rheumatism, cigarettes and cancer (with less or greater accuracy), then a sorting relationship is being assumed between separate realities. The assumption of strong objectivity in classical physics is critical here, since causality describes connections between objects and events clearly independent of the observer. Complementarity is applied more broadly to those deeper realities which, by virtue of the indivisibility of the quantum of action, are inseparable from our interaction with them.

Bohr talks of the complementarity of descriptions in quantum physics and, derivatively, of the complementarity of *psyche* and *physis* in biology, of instinct and reason in human behaviour, of one culture and another culture, of seriousness and humour in art and play, of thought and feeling, love and justice, justice and mercy, and so on. Bohr applies

[47] Letter to Darwin, 20 Aug. 1927, *BSC* : 9, and letter to Pauli, 16 May 1947, *BSC* : 30.

complementarity to aspects, experiences, evidence, pictures, and, especially, phenomena.[48] 'It was the *universal* signficance of the role of complementarity which Bohr came to emphasize', observes Holton.[49] The relationship between complementarity in quantum physics and complementarity in other matters is brought about not so much by the influence of the quantum of action, however, as by the impact of Bohr's epistemological lesson, which will be discussed in the following chapter.

Can any further precision be given to Bohr's usage of 'complementarity'? Weizsäcker fostered a distinction between 'circular' and 'parallel' complementarity, noting that at times Bohr speaks of the complementarity of terms drawn from the same 'genealogy'. For example, 'wave' and 'particle' are both taken from the vocabulary of classical physics and hence are said to be in parallel complementarity. The same might be said of the complementarity of measurements of position and momentum. On the other hand, Bohr's Correspondence Principle evolved into a statement of the complementarity of classical and quantum physics: 'the complementarity between the description of nature through classical concepts and through the [wave-] function . . . is no "parallel complementarity",' says Weizsäcker, 'perhaps one could best describe it as a "circle of knowledge" '.[50] 'Circular' complementarity, therefore, referred to the complementarity of terms drawn from apparently alternative, and perhaps mutually exclusive, frameworks.

It is true that there is a paradoxical relationship between the quantum and the classical approaches: one supplants the other, and yet quantum physics requires classical terminology in order to proceed. It is also true that Bohr delighted in this paradox: 'It is almost the essence of an experiment that the observations can be described with the concepts of classical physics. That is the whole paradox of quantum theory.'[51] In as much as Bohr accepts this holism or hermeneutic circle, Weizsäcker's distinction remains useful, but perhaps he was wrong to suggest that

[48] For these various uses of 'complementarity', see respectively *APHK*, pp. 11, 27, 76, 30, 79, 101, 93; N. Bohr, 'Physical Science and the Study of Religions', in *Studia Orientalia Joanni Pedersen* (Copenhagen, Munksgaard, 1953), 385; *APHK*, pp. 81, 5, 19, 79, 40; *Essays*, pp. 4, 12, 92; *APHK*, pp. 30, 56, 41, 45, 47, 71, 90; and *Essays*, pp. 19, 25, 60.

[49] Holton, 'The Roots of Complementarity', p. 1045.

[50] C. F. von Weizsäcker, 'Komplementarität und Logik', *Die Naturwissenschaften* 42 (1955), 525 f.; see also his *The Unity of Nature* (New York, Farrar Strauss Giroux, 1980), 183.

[51] Bohr, quoted by W. Heisenberg in *Physics and Beyond* (London, Allen and Unwin, 1971), 129; see also *APHK*, pp. 36 f.

complementarity was sometimes only parallel and not circular. Bohr, for his part, rejected Weizsäcker's suggestions as unnecessarily complicating the issue.[52] While I find Weizsäcker's distinction helpful in clarifying what 'complementarity' might entail, it is certainly not essential to an understanding of Bohr's use of the term.

A second distinction, this time between 'strong' and 'weak' complementarity, has been proposed by M. Drieschner. It is quite wrong, he points out, to speak of an architect's ground-plan and elevation-plan as being complementary in Bohr's sense of the word. Such plans are in no way rival or mutually exclusive descriptions, even if they portray different aspects of the one entity. They are, says Drieschner, complementary descriptions in the 'weak' sense of the term. Bohr, on the other hand, wants to see the term in much stronger way: 'wave' and 'particle' are mutually exclusive, just as classical continuity and quantum discontinuity are. One mode of description cannot be reduced to the other. It is because complementarity is paradoxical that it has to be 'strong', and vice versa.[53]

The fullest account of human experience is that which embraces all possible experience. Because the complementarity account makes it possible for us to talk about observations at or beyond the limits of causal certitude, Bohr regards it as providing the possibility of a more exhaustive description. Einstein, for his part, regarded Bohr's approach as a sell-out, a soft option, and reprehensible. In 1926 and 1927 he expressed his instincts in terms like these:

> Quantum mechanics is very impressive. But an inner voice tells me that it is not yet the real thing. The theory produces a good deal but hardly brings us closer to the secret of the Old One. I am at all events convinced that *He* does not play dice . . .
>
> It is only in the quantum theory that Newton's differential method becomes inadequate, and indeed strict causality fails us. But the last word has not yet been said. May the spirit of Newton's method give us the power to restore unison between physical reality and the profoundest characteristic of Newton's teaching—strict causality.[54]

As we shall discuss in Chapter 4, Einstein based his opposition partly on instinct, and partly on a physics of the possibility of 'hidden

[52] See Jammer, *The Philosophy of Quantum Mechanics*, p. 90, with reference to Bohr's letter to Weizsäcker on 5 Mar. 1956.

[53] See M. Drieschner, *Voraussage—Wahrscheinlichkeit—Objekt: Über die begrifflichen Grundlagen der Quantenmechanik* (Berlin, Springer, 1979), 152.

[54] See A. Pais, *'Subtle is the Lord . . .': The Science and Life of Albert Einstein* (Oxford University Press, Oxford, 1982), 15, 443.

variables' (though this was not Einstein's term) which connected the otherwise 'individual' quantum events. If hidden variables did exist, despite our ignorance of them, then a statistical and causal treatment of quantum events would be justified. The need for complementarity would disappear.

'Complementarity' is intended by Bohr to connote a complete framework for intelligent description and unambiguous communication of those aspects of our experience where mutually exclusive and apparently contradictory concepts are employed. It thus applies particularly to events at the bounds of our awareness, the deep truths, where the ordinary univocity of descriptive concepts no longer applies. Because the possibility of further synthesis is out of the question, complementarity is not to be confused with dialectic. In Bohr's Correspondence Principle, however, we find a way of using the benefits of different conceptual frameworks.

2.8. *Correspondence and the Classical*

'Correspondence' and 'classical' are terms frequently used by Bohr and almost as frequently misunderstood by his critics. We have already noted that the first version of the Correspondence Principle was a statement of the asymptotic conformity of results obtained from classical and quantum calculations for higher quantum numbers. This rule of thumb for comparing calculations was then related to a more far-reaching generalization about the use of descriptive concepts. The revised form of the Correspondence Principle, that is, was also connected with the conditions governing our use of concepts drawn from everyday experience. This refinement was expressed by Bohr in the late 1920s: 'the necessity of making extensive use, nevertheless, of the classical concepts, upon which depends ultimately the interpretation of all experience, gave rise to the formulation of the so-called correspondence principle which expresses our endeavours to utilize all the classical concepts by giving them a suitable quantum-theoretical re-interpretation'.[55]

Bohr was well aware of two stages in the development of his notion of correspondence: he moved from the correlation of two theoretically incompatible ways of calculating to a consideration of the necessary use of classical terminology in quantum physics. And by 'classical concepts' he meant not just Newton's and Maxwell's concepts and so on, but also

[55] *ATDN*, p. 8.

all the descriptive concepts drawn from ordinary demonstrable experience. Correspondence, like complementarity, connotes a rational generalization and a framework for unambiguous communication.

The earlier version of the Correspondence Principle is clearly enunciated in Bohr's essays on *The Theory of Spectra and Atomic Constitution*. In 1920 he wrote:

> there is found . . . to exist a far-reaching *correspondence* between the various types of possible transitions between the stationary states [i.e. quantum mechanics] on the one hand and the various harmonic components of the motion [i.e. classical mechanics] on the other. This correspondence is of such a nature, that the present theory of spectra is in a certain sense to be regarded as a rational generalization of the ordinary theory of radiation.
>
> This correspondence between the frequencies determined by the two methods must have a deeper significance and we are led to anticipate that it will also apply to the intensities. This is equivalent to the statement that, when the quantum numbers are large, the relative probability of a particular transition is connected in a simple manner with the amplitude of the corresponding harmonic component in the motion.[56]

In 1921, a year later, Bohr indicated that this device for utilizing results from different methods may also be related to the much broader epistemological issues of reference and meaning: 'quantum theory may . . . be regarded as a rational generalization of our ordinary conceptions.'[57] Because Bohr always thought it important for theory to be related to reality, it follows that the correspondence between theories should also be duplicated in a broader correspondence between the descriptive terms entailed in such theories. Classical terms thus receive a 're-interpretation' in quantum physics.

In his sparse notes for the Gifford Lectures of 1949, Bohr refers to the 'correspondence principle and its early applications'.[58] What he had in mind, I propose, is this first version, which related quantum and classical calculations through a forced marriage of incompatible theories. In later references to the Correspondence Principle, however, Bohr describes it as expressing 'our endeavours to utilize all the classical concepts by giving them a suitable quantum-mechanical re-interpretation'. From the 1930s onwards, Bohr's explicit appeals to the Correspondence Principle become fewer and fewer, but his insistence on its broader interpretation remains prominent: 'Only with the help of

56 *TSAC*, pp. 24, 27. See also *NBCW* 3, pp. 246, 249.
57 Ibid., p. 81.
58 See Bohr's notes dated 3 July 1949 in *MSS*: 19, p. 5.

classical ideas', he declares over and over again, 'is it possible to ascribe an unambiguous meaning to the results of observation.'[59] The arguments behind such assertions, which are related to the conditions for re-identifying particulars in our ordinary experience, will be put on one side for a moment. It is sufficient here to clarify the extent to which Bohr pushes his notion of correspondence and to observe that this surface feature of his interpretation has deeper significance.

Our immediate task must be to show the range of meaning Bohr gives to the term 'classical'. When he uses 'classical ideas', 'classical terms', and 'classical concepts', he has been taken to be referring to 'the scientific concepts which have been used since the Renaissance', as Feyerabend once put it.[60] There is some justification in this, in as much as Bohr speaks of 'electrodynamics', 'motion', 'space-time', 'causality', 'mechanics', and 'laws' as 'classical '.[61] Even in his very late writings he continued to speak of 'the concepts of classical physics'.[62] All the same, it is indisputable that Bohr meant much more than 'post-Renaissance scientific concepts' here. One can, moreover, observe an evolution in his thought: in his writings the emphasis shifts from 'classical concepts' to 'concepts of our ordinary, everyday world'; and such a shift runs directly parallel to the revision of the Correspondence Principle itself. Let us now consider the evidence for this.

In a note which Bohr made in 1930, he described classical physical theories in quite a narrow sense, yet one which left open a much broader possibility: 'By classical physical theories we mean the usual mechanics and electrodynamics which have shown in a wonderful way how to explain ordinary phenomena; these theories are tied very closely to our ordinary attitudes to nature.'[63] More telling still, in notes from 1957 Bohr crossed out the words 'classically defined' and replaced them with 'communicable in ordinary language'.[64] In comparing the various passages in which Bohr states the revised version of the Correspondence Principle, it becomes apparent that he uses the following interchangeably: 'classical terms', 'everyday concepts, perhaps refined by

59 *ATDN*, p. 17. See also pp. 53, 77, 94; *APHK*, pp. 26, 39, 72, 88; and *Essays*, pp. 3, 12, 24.

60 P. Feyerabend, 'Complementarity', *Aristotelian Society Supplement* 32 (1958), 86. In fairness to Feyerabend, it should be noted that in later writings he expanded this narrow interpretation.

61 See *TSAC*, p. v; *ATDN*, pp. 54, 55, 111, 113; and *APHK*, p. 17.

62 *Essays*, p. 3.

63 See note entitled 'Kvanteteorien og de klassiske fysiske Teorier', dated 11 Feb. 1930, *MSS* : 12, p. 1, from the Danish.

64 'Position and Terminology in Atomic Physics', dated 11 Mar. 1957, *MSS* : 22, p. 2.

the terminology of classical physics', 'plain language, suitably supplemented by technical physical terminology', 'the language developed for orientation in our surroundings', 'plain language, suitably refined by our usual physical terminology', 'plain language', and 'common language, suitably refined by the terminology of classical physics'.[65]

Not only does this evidence refute views similar to those once expressed by Feyerabend, it also suggests a more succinct definition of what Bohr means by 'classical': classical concepts are those which apply univocally in the macroscopic world, but which cannot be applied in precisely the same way in the domain of quantum physics. The disjunction in usage is justified because the quantum account is a discontinuous one, whereas classical physics entails a continuity of description. Classical concepts are the more familiar concepts arising from within the everyday world of experience with its usual assumptions of strong objectivity and space-time causal frameworks. They are therefore not necessarily applicable in the same way at or beyond the bounds of ordinary human experience. Bohr himself offers a clue to this negative definition when he speaks of 'classical concepts, neglecting the quantum of action' and states: 'the necessity of basing the description of the properties and manipulation of the measuring instrument on purely classical ideas implies the neglect of all quantum effects in that description'.[66]

One is tempted to declare that classical concepts are simply non-quantum concepts. The problem with this, however, is that there are no such 'quantum concepts' which can be drawn from observation of quantum objects themselves, since a sharp separation between object and observing system is rendered impossible by the indeterminacy relations. This was Bohr's dilemma: in talking about quantum events, which are only observed through interaction with macroscopic apparatus, all the reports are made in classical terms—wave, particle, and so forth. At the quantum level, however, we are entitled to use the

[65] See respectively *APHK*, pp. 26, 72, 88, and *Essays*, pp. 3, 12, 24. An anonymous reader has commented that 'it is not irrelevant here that in the post-war period Denmark came under the spell of ordinary language analysis, and this movement attempted (like many others) to appropriate Bohr as one of their own (cf. Aage Peterson's book on this). They were wrong, of course, but I suspect Bohr's changed terminology has something to do with this influence.'

[66] See *ATDN*, pp. 94 f., and N. Bohr, 'The Causality Problem in Atomic Physics', in *New Theories in Physics* (Paris, International Institute of Intellectual Collaboration, 1939), 19. See also *APHK*, pp. 33, 85.

terms in a different way. Either the terms can be applied analogically rather than univocally (if we think we are applying them to the quantum realm itself, which was not Bohr's way), or they are only applied to the interaction between the quantum system and the classical measuring apparatus. This is part of the 'rational generalization' behind the principles of complementarity and correspondence.[67]

As far as I can tell, Bohr does not explain why he dropped the term 'correspondence' from his later essays. Perhaps it was because the Correspondence Principle is built into quantum mechanics. It also seems likely that he thought correspondence and complementarity expressed much the same point. Certainly, he was able to refine the content of the Correspondence Principle in such a way that the force of both the original and revised versions was not lost, and, secondly, so that the contrast between classical and quantum outlooks remained starkly stated. In 1954 he wrote: 'quantum mechanics presents a consistent generalization of deterministic mechanical description which it embraces as an asymptotic limit in the case of physical phenomena on a scale sufficiently large to allow the neglect of the quantum of action'.[68] In other words, just as statistical determinism is a more general treatment than strict determinism, so also quantum physics is a generalization of classical physics, to which it converges for higher quantum numbers where the effect of the quantum of action becomes negligible.

Correspondence and complementarity were the slogans of Bohr's campaign. His interpretation resolved the dilemma of the fundamental disparity between classical and quantum physics by grasping both horns at once and by explaining why this was permissible. The justification for this, ultimately, rests on reflections on the nature of our language and concepts, on how we learn our words from the everyday world, and vice versa. By prescribing this solution, Bohr was able to restore some order into the relationship between quantum and classical physics. The price, however, was high: the abandonment of the ambitions of classical physics for a strictly deterministic mechanical account of physical reality.

Having dealt with the main surface features of Bohr's interpretation of quantum theory, we are almost in a position to consider his underlying arguments. Before doing so, however, it is important to examine

[67] Bohr uses 'rational generalization' often in order to describe his coupling of the quantum and the classical through complementarity and correspondence: see *ATDN*, pp. 70, 87, and *APHK*, pp. 85, 90, 100.

[68] *APHK*, pp. 73 f.

Bohr's resolution of the problem of measurement in quantum physics. This will also allow us to clarify his notions of 'phenomena' and 'individuality'.

2.9. *Measurement, Phenomena, and Individuality*

There was little discussion in classical physics of the nature of observation and measurement. The impact of quantum theory, and in particular of Heisenberg's indeterminacy relations, raised a serious challenge to the rather naïve assumption that what was observed in classical physics was the state of the object itself.

The problem of observation in quantum physics was restated as the more specific problem of measurement. Bohr's views on the issue are as illuminating now as they were controversial then. His apparently doctrinaire approach has its deeper reasons. A thorough formal theory of measurement, either in classical or quantum physics was, to his mind, out of the question, since any axiomatization would depend on primitive descriptive concepts drawn from the language justified by ordinary experience. In our discussion of measurement, also, we can provide further details of his conception of the role of subject and object in human knowing.

Bohr's provocative tendency, especially in earlier writings, to 'emphasize the subjective character of all experience'[69] brought his entire interpretation of quantum theory into peril. Bunge and Popper, for example, focus on this aspect of Bohr's thought and reject his position outright as purely subjectivist and positivist. That they have neglected the context of Bohr's remarks and misconstrued his intent has been pointed out by Feyerabend and Audi.[70]

Bohr never intends to suggest that observation somehow constitutes reality. He is making another point: in quantum physics a different kind of objectivity prevails, one which demands recognition not only of the external reality, but also of the descriptive frameworks employed by the

[69] *ATDN*, p. 1.

[70] See Bunge and Popper in M. Bunge (ed.), *Quantum Theory and Reality*, Studies in the Foundations, Methodology, and Philosophy of Science, 2 (New York, Springer, 1967), and M. Bunge, 'Strife about Complementarity', *British Journal for the Philosophy of Science* 6 (1955), 1-12. For replies to these arguments, see P. K. Feyerabend, 'On a Recent Critique of Complementarity', *Philosophy of Science* 35 (1968), 309-31, and 36 (1969), 82-195 (reprinted in his *Realism, Rationalism and Scientific Method* (Cambridge, Cambridge University Press, 1981), 247-97); and M. Audi, *The Interpretation of Quantum Physics* (Chicago, University of Chicago Press, 1973).

observer. The full context of the quotation above gives Bohr's position:

> . . . all new experience makes its appearance within the frame of our customary points of view and forms of perception. . . . In physics, where our problem consists in the co-ordination of our experience of the external world, the question of the nature of our forms of perception will generally be less acute than it is in psychology, where it is our own mental activity which is the object under investigation. Yet occasionally just this 'objectivity' of physical observations becomes particularly suited to emphasize the subjective character of all experience.

To say that experience has a subjective character is not to say that it is *purely* subjective.

Bohr was reacting here against the ideal of strong objectivity inherent in the programme of classical physics. Even in his early remarks on the nature of measurement, however, he does not deny the possibility and importance of public points of reference: 'In tracing observations back to our sensations, once more regard has to be taken to the quantum postulate in connection with the perception of the agency of observation, be it through its direct action upon the eye or by means of suitable auxiliaries such as photographic plates . . .'[71]. If there remains some ambiguity in this, in that 'eye' and 'photographic plate' may not seem to be equally public, it should be noted that it is the *interpretation* of visible sense-data which is 'private', rather than the eye itself. And if it is still not clear which means of observation Bohr regards as central, in later writings he strongly rebuffs the possibility of being accused of subjectivism:

> It is also essential to remember that all unambiguous information concerning atomic objects is derived from the permanent marks . . . left on the bodies which define the experimental conditions. . . . The description of atomic phenomena has in these respects a perfectly objective character, in the sense that no explicit reference is made to any individual observer . . . As regards all such points, the observation problem of quantum physics in no way differs from the classical physical approach.[72]

It is to statements like these that Feyerabend and Audi refer in their defence of Bohr. Their detailed criticisms of Bunge and Popper need not be duplicated here. Instead we should note how Bohr has removed the measurement problem from the context of a debate between quantum and classical physics: he regards observation in both realms as

[71] *ATDN*, p. 67.
[72] *Essays*, p. 3.

entailing the same steps and the same problems. Bohr's interest is a more fundamental one: the epistemology of the actor and the spectator.

The problem of measurement in quantum mechanics is, for Bohr, a particular instance of a universal problem. It highlights the issue of the relationship between language and fact, word and world, theory and observation. If one accepts the 'indivisibility' of the quantum, and if one takes the point about the 'elusiveness' of the ultimate subject (Bohr's spectator-actor dilemma), then both are concerned with knowledge at the perimeter of human experience. Note that 'subject' here refers sometimes to the 'observing system', sometimes to the 'self' of consciousness, and sometimes to both. In quantum physics, the key factor is the discontinuity which the quantum forces us to accept in our account of microphysical events and processes. The finite magnitude of the quantum of action, says Bohr repeatedly, prevents altogether a sharp distinction being made between the object of measurement and the agency by which it is observed:

> the indivisibility of the quantum of action demands that, when any individual result of measurement is interpreted in terms of classical conceptions, a certain amount of latitude be allowed in our account of the mutual action between the object and the means of observation. This implies that a subsequent measurement to a certain degree deprives the information given by a previous measurement of its significance for predicting the future course of the phenomena [here meaning 'objects of measurement']. Obviously these facts not only set a limit to the *extent* of the information obtainable by measurements, but they also set a limit to the meaning which we may attribute to such information.[73]

As a result, Bohr occasionally argued for the 'fundamental limitation, met with in atomic physics, of the objective existence of phenomena ['objects of experience'] independent of their means of observation'. Again, our purpose in the description of nature, as he put it in his earlier writings, is 'not to disclose the real essence of the phenomena but only to track down, so far as it is possible, relations between the manifold aspects of our experience'.[74] It may appear from this that Bohr espoused the idea that physics was concerned only with an account of measurement-observations, and not with reality. This is only half the story, however, and Bohr strongly objected to any such accusations. He was disconsolate when his views were openly accepted by a group of

[73] *ATDN*, p. 18, Bohr's emphasis. See also pp. 5, 11.

[74] See respectively *APHK*, p. 7, and *ATDN*, p. 18; see also *APHK*, pp. 25, 69, 99, and 'The Causality Problem in Atomic Physics', p. 24.

positivist philosophers.[75] He also disapproved of the phrase, 'creation of physical attributes to atomic objects by measurements'.[76] Bohr did not want to stress either the role of the subject alone, or that of the independent object. Rather, he wanted to insist on the wholeness of the interaction between observer and observed.

Realizing the dangers of ambiguity on this point, Bohr later refined his terminology and clarified what he meant by 'phenomenon'. After 1939 he employed 'the word "phenomenon" to refer only to observation obtained under circumstances whose description includes an account of the whole experimental arrangement'.[77] The term was not intended to signify the interpreted appearance of the object of experience itself. Nor was Bohr trying to follow the Kantian distinction between the thing-in-itself and our perception of it. If one wanted to talk about such 'things', then they were, as Weizsäcker put it, to be found *in* the phenomena rather than behind it.

To understand Bohr here, one has to get away from a theory of knowledge in which the mind is presumed to mirror reality. He would support the kinds of argument mustered by Rorty, for example, in the criticism of foundationalist epistemologies. On the other hand, Bohr would certainly maintain that we can know an external reality, but the knowing is more by participation than by some sort of abstraction at a distance. Reference to the whole experimental arrangement thus takes into account the manner of the observer's participation rather than proposing the creation of attributes of atomic reality.

Bohr also set out to clarify the meaning of the terms 'observation' and 'measurement'. Both were equivalently described as entailing 'unambiguous information . . . derived from the permanent marks . . . left on the bodies which define the experimental conditions'. Measurement and observation were thus interconnected with Bohr's notion of phenomena. Because of the quantum discontinuity, no precise demarcation could be drawn between 'object' and 'measuring instrument': 'The essential wholeness of a proper quantum phenomenon finds indeed logical expression in the circumstance that any attempt at its well-defined subdivision would require a change in the experimental arrangement incompatible with the appearance of the phenomenon itself'. Thus we are 'dealing with phenomena where no sharp distinction can be made between the behaviour of the objects themselves and their interaction with the measuring instruments'. If we

[75] See Weizsäcker, *The Unity of Nature*, pp. 185 ff.

[76] See *APHK*, p. 73, and *Essays*, p. 5.

[77] *APHK*, p. 73.

are to talk of measurement and what we are measuring, then the nature of the experimental arrangement is crucial: 'the unambiguous account of proper quantum phenomena must, in principle, include a description of all relevant features of the experimental arrangement'.[78]

In Bohr's later usage of 'phenomenon', therefore, the term is the equivalent of 'observation', which he uses to denote a physical interaction between a quantum system and the instruments used for measurement and observation. This clarification of usage occurred as a result of the debates between Bohr and Einstein, and in particular as a result of the Einstein-Podolsky-Rosen proposal, topics which will be discussed in Chapter 4.

Bohr's approach is more epistemological than ontological, and his neglect of ontological questions leaves scope for much criticism. But Bohr's talk of 'phenomena' at the quantum level has its roots in the reality of the everyday, public, macroscopic world as we classically know it. Bohr wants to accept such knowledge not only as valid, but also as open to generalization. In this sense, his is also a broad realism. While his focus on 'individuality' seems to imply a subjective realism about quantum events, this is complemented, in the strong sense of the term, by his unquestioning acceptance of macroscopic reality as part of 'nature'.

The term 'individuality' is meant to denote the 'wholeness' of the subject-object interaction, rather than any splendid isolation of the subject. Indeed, in successive drafts of one of his essays, Bohr replaces 'wholeness' in turn by 'indivisibility' and 'unity'. In parallel passages he uses 'atomicity', 'discontinuity', and 'individuality'.[79] When we apply macroscopic instruments to observe quantum events, as we must, two conceptual frameworks overlap in that wholeness. The subjectivity, therefore, refers not only to the knower's contribution, but to the individuality of quantum events when the observer intervenes.

Measurement, given the introduction of quantum theory, differs from the classical account in one important respect. On the classical scale, because the quantum discontinuity of change can be neglected or accounted for, measurement is taken to be of the mechanical system itself, independent of the measuring apparatus. One observes the state of the system, and the classical dynamics and mechanics offer a theoretical description of the state of that system both before and after

[78] See respectively *Essays*, p. 3, *APHK*, pp. 72 and 62, and *Essays*, p. 4.

[79] See p. 7 of the drafts of 'Mathematics and Natural Philosophy', *MSS* : 20, and Stolzenburg, *Die Entwicklung des Bohrschen Komplementaritätsgedanken*, p. 157.

measurement, provided the system remains closed. Although measurement always requires an interaction between the observed system and the measuring system, such an interaction only becomes disruptive and critical at the quantum level. The quantum formalism does not describe the unobserved state of the system, but the state of the system in interaction with the measuring system. The individuality of the measuring process refers not only to the indivisibility of the interaction between the two systems, but also to the uniqueness of each measurement. This is because measurement itself is the event whereby a description of the character of the interaction with the quantum process can be conceived, and subsequent measurement is a new conception rather than continuous with the earlier measurement.

The difference between Bohr and Einstein on this point is not that one is a positivist and the other a realist. Einstein suggested that a statistical account of quantum events would remove the problem of individuality; Bohr, insisting on such individuality, must therefore provide a different interpretation of the reality of quantum processes.

Problems about realism will also be discussed in Chapter 4, in the context of the Bohr-Einstein debates. So far we have completed a preliminary survey of some of the prominent features of Bohr's interpretation of quantum theory. Among the most important of Bohr's convictions is his assertion that, from the point of view of physics, there is a necessary lower bound to the possibility of fitting our experiences into the usual conceptual frameworks, and hence also a limit to the application of a causal, spatio-temporal account of reality. We now turn to the underlying arguments which support his interpretation, the transcendental defence of these convictions.

3

Bohr's Transcendental Philosophy

> I am occupied by trying to trace the philosophical aspect a bit further.
>
> Niels Bohr

3.1. *Bohr and Philosophy*

'It may not be too much an exaggeration', according to Jørgen Kalckar, 'to describe Bohr as a born philosopher of nature, who found in physics a marvellously powerful instrument for probing into the foundations of human knowledge and man's description of the world.'[1] If Bohr was 'primarily a philosopher', as Heisenberg put it, what is his place among philosophers? Where does he stand? Two facets of Bohr's disposition make it difficult to answer these questions. He seemed deliberately to take up a stance at some distance from the usual forum for scholarly philosophical debate, and, secondly, his concern for precision of expression produced constant revisions of his usual arguments, rendering them even more elusive.

'It is not my intention', he once wrote, 'to enter into an academic philosophical discourse for which I would hardly possess the required scholarship.'[2] Bohr's correspondence in the late 1920s, however, reveals his intention to pursue his own philosophical path. In March 1928, writing about his work on the interpretation of the 'new' quantum theory, Bohr confided in Darwin: 'I do not know what physicists will think about it, at any rate I have put a lot of work into the formation of my views. At present I am occupied by trying to trace the philosophical aspect a bit further . . .'.[3] Similarly, in the same year, Bohr told Høffding that 'it is indeed the purely epistemological side of the analysis of concepts that I have preoccupied myself with in my work'.[4] This consideration of the conditions which apply to the unambiguous use of concepts and to our conceptual frameworks was to appear in

[1] J. Kalckar, 'General Introduction', *NBCW* 6, p. xvii.
[2] *APHK*, p. 67.
[3] Letter to Darwin, 26 Mar. 1928, *BSC* : 9.
[4] Letter to Høffding, 1 Aug. 1928. *BSC* : 12.

Bohr's later work as a simple epistemological lesson. After what may seem a rather roundabout approach, we will return to analyse this fundamental and transcendental consideration as a key to Bohr's interpretation.

In 1929, responding to Bohr's recent efforts to provide an interpretation of the new quantum formalism, Pauli congratulated him for having omitted all physics and for having concentrated purely on philosophy. Pauli then went on to suggest that evaluation should be left to 'the professional philosophers'.[5] Bohr, ever a free spirit, was to protect himself from such potential criticism by saying: 'With all appreciation and admiration for the refinement by which such problems [of knowledge] have been discussed through the ages in the schools of sophistic, empirical or realist philosophy . . . such endeavours are to my mind not directly connected with our task.'[6] He felt that 'philosophers were very odd people who really were lost',[7] who could not understand him, and among whom he really did not wish to be numbered. 'He certainly did not think highly of the teachings of "professional" philosophers', Kalckar recalls.[8] Perhaps Léon Rosenfeld, a close colleague, has put his finger on the sensitive point in Bohr's make-up: he 'intensely disliked the idea of having a label stuck on him'.[9]

The second factor that makes Bohr's philosophical position hard to determine is, perhaps surprisingly, his concern for precision of expression: because his words were ever on the move, the reader is always trying to catch up with him. J. Rud Nielsen, for example, tells the harrowing story of Bohr's demands for nine different sets of page-proofs from an editor before he was satisfied with what he had wrought.[10] Bohr himself acknowledges that his work displays a gradual evolution and refinement of terminology, with the consequence that variations and inconsistencies appear across the span of fifty years of writing.[11] Indeed, the roundabout approach here corresponds to Bohr's own often baffling way of making himself clear. Those philosophers used to more direct assaults on clarity will, unfortunately, remain baffled.

[5] Letter from Pauli, 17 July 1929, *BSC* : 14; see *NBCW* 6, p. 447.

[6] See Bohr's notes for 'The Unity of Knowledge', dated 29 Mar. 1954, *MSS* : 21.

[7] See the interview with Kuhn on the day before Bohr's death, *AHQP*, interview 5, p. 3 of transcript.

[8] Kalckar, 'General Introduction', *NBCW* 6, p. xx.

[9] Letter from Rosenfeld to Stapp, in H. P. Stapp, 'The Copenhagen Interpretation', *American Journal of Physics* 40 (1972), 1115.

[10] J. Rud Nielsen, 'Memories of Niels Bohr', *Physics Today* 16 (1963), 25.

[11] See Bohr's Introduction to *APHK*, p. v.

Bohr's difficulties in saying what he wanted to say, combined with his attitude towards classical physics, appears to some critics to be a kind of mystery-mongering, casting a veil over his work. Heisenberg concedes that 'it may be a point in the Copenhagen interpretation that its language has a certain degree of vagueness, and I doubt whether it can become clearer by trying to avoid this vagueness'.[12] Such an observation foreshadows later discussion of the nature of transcendental claims, where the *articulation* of necessary conditions is always open to alternative forms of expression. In Bohr's writings, moreover, it may also have something to do with his very inventive and occasionally obscure use of English. All the same, it is important to distinguish between what is cryptic, repetitious, and enigmatic on the one hand, and what is circular or holist on the other. Too often the former has been allowed to overshadow the latter.

If Bohr's approach is not recognized as more circular (that is, as acknowledging the hermeneutic circle of language and reality) than axiomatic or architectonic, then it will remain baffling. The circularity is not a vicious one, a boot-strapping operation, but self-referential. Bohr asks us to discover what is constitutive of our unambiguous communication of experience by reflecting on what we do when we attempt to communicate experience unambiguously. He does not choose to begin with a foundational epistemology resting on an absolute and permanent bedrock of privileged word-world relationships.

Attempts to identify Bohr with this or that philosopher have frequently been made. His path has been variously traced back to Kierkegaard, Høffding, Mach, James, and especially Kant.[13] Such forays might be excusable, but the spoils have definitely been poor.[14] While passing reference can occasionally be found in Bohr's notes to the ancient Greeks, Spinoza, Descartes, Hume, Berkeley, Boscovich,

[12] Letter to Stapp, in Stapp, 'The Copenhagen Interpretation', p. 1113.

[13] See, for example, M. Jammer, *The Conceptual Development of Quantum Physics* (New York, McGraw Hill, 1966), 173; G. J. Holton, 'The Roots of Complementarity', *Daedalus* 99b (1970), 1041 ff.; S. L. Jaki, *The Origins of Science and the Science of its Origins* (Edinburgh, Scottish Academic Press, 1978), 200 ff.; C. A. Hooker, 'The Nature of Quantum-Mechanical Reality', in R. G. Colodny (ed.), *Paradigms and Paradoxes* (Pittsburgh, University of Pittsburgh Series in the Philosophy of Science 5, 1972), 135; K. M. Meyer-Abich, *Korrespondenz, Individualität und Komplementarität* (Wiesbaden, Steiner, 1965), 133-40.

[14] See D. Favrholdt, 'Niels Bohr and Danish Philosophy', *Danish Yearbook of Philosophy* 13 (1976), 206-20; Kalckar, 'General Introduction', *NBCW* 6, p. xxii; and K. Stolzenburg, *Die Entwicklung des Bohrschen Komplementaritätsgedankens in den Jahren 1924 bis 1929* (Stuttgart, Ph. D. thesis, 1977), 151 f.

Mach, and even Kant,[15] there is no evidence that these philosophers ever had any direct influence on Bohr's work. There have also been endeavours to categorize Bohr as a pragmatist, idealist, positivist, and so on. Hooker, testing the fit of no less than seven philosophical guises on Bohr, acknowledges the risks of mistaken identification.[16] In so far as one can describe Bohr as a philosopher, one must concede at the outset that his philosophy is *sui generis*.

There is, all the same, an underlying philosophical position. Bohr and his colleagues were convinced of the philosophical character of his work and of the significance of his contribution to philosophy in general. Describing his approach to a paper for the Planck Jubilee in 1929, Bohr declared to Pauli: 'it got much too long and took so much time that at the last moment I had to leave all physics out of my article and stick to pure philosophy, and even that only by way of allusions'.[17] In later years these allusions were sharpened, as we will see, into an 'epistemological lesson'. Bohr will claim to have discovered something 'hitherto unnoticed' about the presuppositions of human knowledge.[18]

The centrality of philosophy to Bohr's interpretation of quantum theory has been strongly stated not only by Heisenberg and Weizsäcker, but also by later collaborators like Rosenfeld, Petersen, and Kalckar.[19] Despite these testimonies, a critical account of the character of this philosophy has yet to be established. My immediate concern here, then, is not with the history of the development of Bohr's thought, nor, for the moment, with the validity of his interpretation, but with the nature of his philosophical approach. If I term it 'transcendental', my intention is not to put a label on Bohr so much as to identify how his approach is related to the mainstream of philosophy.

Questions about the formation of concepts were always of interest to Bohr. In his opinion, physics provided a special impetus to philosophy: 'the opportunities which time and again it has offered for examination

[15] See notes dated 5 Aug. 1953 (*MSS* : 20), 3 Sept. 1954 (*MSS* : 21), 27 Oct. 1941 and 6 Feb. 1942 (*MSS* : 16), and 6 Aug. 1953 (*MSS* : 20). Bohr actually gave speeches at occasions commemorating both Mach and Boscovich, but made no significant reference to either thinker: see *MSS* : 15, and *MSS* : 23. Kant is mentioned by name on notes dated 6 Feb. 1942 (*MSS* : 16), and preceding those dated 27 Oct. 1941 (*MSS* : 20).

[16] See Hooker, 'The Nature of Quantum-Mechanical Reality', pp. 160 ff., 170, 193, and 206.

[17] Letter to Pauli, 1 July 1929, *BSC* : 14. See *NBCW* 6, p. 443.

[18] *APHK*, p. 91; see also 'hitherto disregarded' on pp. 18 f.

[19] See L. Rosenfeld's 'Biographical Sketch', in *NBCW* 1, p. xvii; and A. Petersen, 'The Philosophy of Niels Bohr', *Bulletin of the Atomic Scientists* 19 (1963), 8-14, and *Quantum Physics and the Philosophical Tradition* (London, MIT Press, 1968).

and refinement of our conceptual tools'.[20] In the remaining sections of this chapter we shall look at his reflections on the role of concepts through his early correspondence and his later writings. The period between 1930 and 1937 is given less emphasis for two reasons: first, the collection of essays that appeared in 1934 lacks the force and subtlety of Bohr's idiosyncratic English, since the translation was undertaken by some of his students; secondly, in the debate with Einstein which dominated discussions of quantum theory between 1930 and 1936, Bohr's attention was directed to a series of thought-experiments in order to meet Einstein's counter-claims on Einstein's own ground. This particular interlude will be dealt with in the following chapter.

We begin the study of Bohr's philosophy with his early correspondence, which contains the origins of his later formulations. The major part of this chapter will then be given over to a critical exegesis of his subsequently published essays. In this exegesis the stages in Bohr's reasoning are identified more as a set of propositions than as a sequential argument, and the final task of this chapter will be to show how, in Bohr's mind, such propositions are neither rhetorical nor pragmatic, but necessarily true.

3.2. *Early Correspondence: Limitations and Conditions*

Kalckar tells us that Bohr thought a great deal about epistemology while still a university student and that, even before he finished his studies, Bohr intended to publish a book. 'As far as we know,' says Kalckar, 'Bohr's youthful philosophical ideas revolved around the problem raised by the movability of the partition between subject and object.'[21]

One of the earliest indications of Bohr's interest in the problems of knowledge and perception occurs in a letter written to his brother Harald in 1910. Commenting on the difficulty he had in harmonizing various sources of elation, Bohr observes that 'sensations, like cognitions, must be arranged in planes that cannot be compared'.[22] Poignantly, as Rosenfeld notes, a similar reference to Reimannian geometry was on Bohr's blackboard at the time of his death. Rosenfeld explains Bohr's insight in the following way: 'The use of words in everyday life must then be subject to the condition that they be kept within the same plane of objectivity; and as soon as we deal with words

[20] *APHK*, p. 1.

[21] Kalckar, 'General Introduction', *NBCW* 6, p. xxiv.

[22] Letter to Harald Bohr, 26 June 1910; see also *NBCW* 1, p. 513, and Rosenfeld's comments, p. xxi.

referring to our own thinking, we are exposed to the danger of gliding onto another plane.'[23] This is well stated, for it draws together Bohr's interests in conditions for the proper use of concepts and in the possibility of concepts being applied in quite different realms only if different frameworks are employed. If a change in our conceptual frameworks occurs, then it cannot be assumed that our concepts can be applied as before. A new framework may be required. Here also, then, is the germ of the notions of correspondence and complementarity.

Rosenfeld could also have stressed the importance of boundary conditions and limitations in Bohr's thinking. In 1914 Bohr wrote on this theme to his patron, the Swedish physicist Carl Oseen, in a letter full of portents: 'the possibility of a comprehensive picture should perhaps not be sought in the generality of the point of view, but rather in the strictest possible limitation of the applicability of the points of view'.[24] In other words, Bohr is emphasizing the importance of considering the bounds to the possibility of forming and applying 'our conceptual tools'. All his subsequent talk about the necessity for using classical terms in the description of quantum events is based on this consideration. And, in the early 1920s, Bohr notes his interest not just in the difficulties of the old quantum theory, but also in the 'inadequacy of our ordinary theoretical conceptions'.[25]

I have already quoted at length from Bohr's letter to Høffding in 1922 about the role of analogies and the application of everyday concepts in the new physics: 'In general,' he wrote, 'and particularly in some new fields of investigation, one must remember the obvious or likely inadequacy of pictures.' In a further passage Bohr raises the related issue of conceptual frameworks, suggesting that, in quantum physics, we will be unable 'to carry through a description in space and time that corresponds to our ordinary sensory images'.[26] These considerations, later to be expressed in terms of 'content' and 'formal frame', will appear as key steps in his defence of his interpretation of the quantum formalism.

Problems associated with the original quantum theory of the atom also contributed to Bohr's thinking about limitations and boundary conditions. In 1924, about the time of the Bohr-Kramers-Slater

[23] L. Rosenfeld, 'Niels Bohr's Contribution to Epistemology', *Physics Today* 16 (1963), 49, 54.

[24] Letter to Oseen, 28 Sept. 1914, *BSC*: 5; see *NBCW* 2, p. 563.

[25] *TSAC*, p. 4.

[26] Letter to Høffding, 22 Sept. 1922, *BSC*: 3, quoted at length above, section 2.6.; see also n. 33.

proposal, Bohr wrote to R. H. Fowler: 'I believe we have here to do with an instructive example of the limitations in the ordinary quantum theory rules for collisions which affords an illustration of the necessity of giving up the strict validity of the general principles of conservation of energy and momentum.'[27] This was written, remember, some years before Heisenberg fashioned the indeterminacy relations which gave so much support to Bohr's intuitions. Once Heisenberg's work had been accepted, Bohr's interest in the constitutive role of limiting conditions to concepts and frameworks became even more intense.

It will be recalled that the indeterminacy relations were established by Heisenberg shortly before Bohr's Como address in 1927, in which he aired the notion of complementarity for the first time. They showed very dramatically how quantum theory came into conflict with the classical deterministic description of physical reality at the atomic level. The fact that Heisenberg at that time associated his discovery with a somewhat positivistic view of physics, pushed Bohr into further refinement of his own point of view.

In the hectic time after Heisenberg's discovery, therefore, several reflections on this covert transcendental theme are to be found in Bohr's correspondence. His focus is on the necessary conditions governing the formation of concepts and the possibility of giving precise description. To Einstein, for example, Bohr wrote in 1927:

It has of course long been recognized how intimately the difficulties of quantum theory are connected with the concepts, or rather with the words that are used in the customary description of nature, and which all have their origin in the classical theories . . . This very circumstance that the limitations of our concepts coincide so closely with the limitations in our possibilities of observation, permits us—as Heisenberg emphasizes—to avoid contradictions . . .

As regards such a pedagogically coloured concept as visualizability, however, it seems to me instructive always to keep in mind how indispensable are the concepts of the continuous field theory in the present stage of science. As long as we only talk about particles and quantum jumps, it is difficult to find a simple presentation of the theory, *which is based on a reference to the limitation in the possibilities of observation.* This is because the uncertainty mentioned is *not only connected to the presence of discontinuities but also to the very impossibility of a detached description in accordance with those properties of material particles in light that find expression in the wave theory.*[28]

[27] Letter to R. H. Fowler, 5 Dec. 1924, *BSC* : 10 (*NBCW* 5, pp. 334 f.); see also the letter to Max Born, 9 Dec. 1924, *BSC* : 9 (*NBCW* 5, p. 300), for a similar remark.

[28] Letter to Einstein, 13 Apr. 1927, *BSC* : 10 (*NBCW* 6, pp. 21 ff., 418 ff.), my emphasis.

This is a passage of the highest significance. Bohr states with astonishing directness that the quantum theory is based on the conditions for the possibility of observation. This is unequivocally a transcendental claim of a general kind. Secondly, Bohr relates the difficulties which result from the introduction of the quantum theory (namely, discontinuity versus continuity) to a general consideration of the possibilities of complete description. Both these themes will be discussed in detail in the exegesis of Bohr's later writings below. Perhaps nowhere else, however, did Bohr show his hand so clearly.

The letter to Einstein is not an isolated instance, an aberration on Bohr's part. In 1928 he wrote to Dirac in a similar frame of mind:

> I think, we can not too strongly emphasize the inadequacy of our ordinary perception when dealing with quantum problems. . . . [*The*] *possibility of some kind of remembrance is of course the necessary condition for making any use of observational results*. . . . In this respect it appears to me that the emphasis on the subjective character of the idea of observation is essential. Indeed I believe that the contrast between this idea and the classical idea of isolated objects is decisive for the limitation which characterizes the use of all classical concepts in the quantum theory.[29]

Here the second sentence constitutes a quite baldly stated transcendental claim and once again discloses Bohr's habitual way of reasoning. The last sentence, furthermore, links the subject-object dilemma to the difficulties faced by physicists in coming to terms with the implications of the quantum theory.

Bohr's letter to Oseen in 1928 also refers to this parallel between the problem of subject-object separation, 'the way of working of our consciousness', and the implications of the quantum postulate for objective description in physics:

> As we already discussed years ago, the difficulty in all philosophy is the circumstance that *the functioning of our consciousness presupposes a requirement* as regards the objectivity of the content, while on the other hand the idea of the subject, of our own ego, forms a part of the content of our consciousness. This is exactly the kind of difficulties of which we got such a clear example in the character of the description of nature required by the essence of the quantum postulate. Far from bemoaning the fact that in atomic physics our usual wishes with respect to the description of nature cannot be fulfilled, *I believe we ought to rejoice at the new lesson concerning the limitation in the human forms of visualization that is implied by the discovery of the quantum of action.*[30]

[29] Letter to Dirac, 24 Mar. 1928, *BSC* : 9 (*NBCW* 6, pp. 45 f.), my emphasis.
[30] Letter to Oseen, 5 Nov. 1928, *BSC* : 14 (*NBCW* 6, pp. 189 f., 431), my emphasis.

Bohr's early letter to his brother in 1910 can now be read as revealing the same concern with boundaries and limiting conditions. But it is in the letters of 1928 which we have quoted, as well as in others to Schrödinger, Fowler, Ehrenfest, Pauli, and Birtwistle, that one sees Bohr giving all his attention to the conditions which constitute the possibility of perception and objectivity.[31] Contemporary non-positivist philosophers discuss these same issues as the problems of the interplay of theory and observation, language and fact, word and world, and so on.

Bohr's assertion that the essence of the quantum postulate implies an unavoidable boundary for the possibilities of definition[32] will be referred to below as the *quantum postulate.* This is a more general statement of the theoretical formula which Bohr introduced as the quantum condition in 1913, the relationship which limited the possible orbits of atomic electrons to a factor of Planck's constant. The general quantum postulate was the well-spring of Bohr's interpretation of the new quantum formalism. From it flows the so-called Correspondence Principle, in its revised version, as well as the notion of complementarity.

Bohr's early attempts to articulate his interpretation at Como and the Solvay Conference in 1927 may not reveal this sequence of argument so clearly, but then Bohr was notorious for beginning sentences which he could not finish. His brother, who was an excellent lecturer, is once supposed to have been asked to explain why his teaching was always so clear and his brother's sometimes so obscure. According to the story, Harald replied: 'Everything which I say follows exactly from what I have just said, but everything which Niels says follows exactly from what he is about to say.' And it is to what Bohr was about to say that we now turn. In his later published writings, the key role of transcendental reflections on limitation and possibility is as apparent as it is in his early correspondence. In particular, we will discover that the 'quantum postulate' or 'quantum condition' is a specific instance of a more general transcendental consideration. This is not to say that Bohr altered his thinking, but only that the stages of his argument become a little more obvious in his later writings.

[31] See the letters to Schrödinger, 23 May 1928, *BSC*: 16 (*NBCW* 6, pp. 48, 464); to Fowler, 19 May 1928, *BSC*: 10; to Ehrenfest, 31 May 1928, *BSC*: 10; to Birtwistle, 3 May and 15 June 1928, *BSC*: 9; and to Pauli, 7 Aug. 1928, *BSC*: 14.

[32] Letter to Schrödinger, 23 May 1928, *BSC*: 16, from the German.

3.3. *Later Writings: A Note on Exegetical Method*

The following sections of this chapter contain the results of a detailed comparison of eleven of Bohr's major essays published between 1937 and 1963, a year after his death. In each of these essays Bohr is concerned with the relationship between the implications of quantum theory and some other aspect of human enquiry, and, in every case, his thinking includes a sequence of four considerations: (*a*) the quantum postulate; (*b*) the conditions for unambiguous communication; (*c*) a prescription for complete description; and (*d*) the deep-going analogy with the subject-object distinction. Bohr also regularly appeals to a fifth consideration: (*e*) the conditions for the ordering of experience.

This last step is separated from the others because its position in the sequence of Bohr's thought varies. In the earlier essays under consideration it appears at the conclusion of his chain of reasoning, whereas in the later essays it is given a distinguished place at the beginning of his argument. I shall demonstrate that (*e*) can be taken as a kind of 'zeroth' assertion: that is, coming at both the beginning and the end of Bohr's chain of reasoning, tying the whole together. Self-referential in character, this step both initiates and completes the circle of his thought. It both precedes, and follows from, the other stages in his argument.

After comparing eleven versions of Bohr's account in eleven different essays, I am convinced of the consistency of the manner in which he establishes his position, despite variations in terminology. It would be tiresome and perhaps counter-productive to include all the excerpts here, so I have confined myself to three representative essays, each separated by a decade: 'Biology and Atomic Physics' (1937); 'Discussion with Einstein on Epistemological Problems in Atomic Physics' (1949) and 'Quantum Physics and Philosophy' (1958). References to parallel passages will chiefly be confined to footnotes.

The following sections, then, assemble the evidence for the stages (*a*) to (*e*) above, and outline further elements in Bohr's argument under each heading. The concluding sections of the chapter contain an assessment of Bohr's argument and an unadorned statement of its key stages.

3.4. (a) *The Quantum Postulate*

Bohr's first move in the earlier essays under consideration is to make the bold claim that the discovery of the universal and elementary quantum of action has brought about an entirely new set of circumstances for the understanding of human enquiry into nature. In as much as the quantum introduces a necessary 'individuality' (or wholeness, unity, atomicity, discontinuity), so also it imposes a limiting condition upon the scope of classical or everyday descriptions and the applicability of the usual conceptual frameworks. In 1937 Bohr wrote:

The existence of the elementary quantum of action expresses, in fact, a new trait of individuality of physical processes which is quite foreign to the classical laws of mechanics and electromagnetism and limits their validity essentially to those phenomena which involve actions large compared to the value of a single quantum . . .

Any attempt to analyse . . . 'individuality' of atomic processes, as conditioned by the quantum of action, will be frustrated by the unavoidable interaction between the atomic objects concerned and the measuring instruments.[33]

Eleven years later, Bohr put forward much the same argument and much the same explanation:

This discovery [of the universal quantum of action], which revealed a feature of atomicity in the laws of nature going far beyond the old doctrine of the limited divisibility of matter, has indeed taught us that the classical theories of physics are idealizations which can be unambiguously applied only in the limit where all actions involved are large compared with the quantum. . . .

[The] individuality of the typical quantum effects finds its proper expression in the circumstance that any attempt of subdividing the phenomena will demand a change in the experimental arrangement . . .[34]

The word 'idealization' may have been derived from the then commonplace view that Newtonian mechanics is an idealization of relativistic mechanics if the speed of light is allowed to approach infinity, but the word also implies more. Flat-earth theory, for example, can only be applied unambiguously on a carefully blinkered scale: to apply flat-earth theory on a grand scale would be to idealize the experience of a few flat hectares. Similarly, while classical physics applies successfully within certain limits, it is an idealization to think that it applies universally, especially on the sub-atomic scale. If one accepts this limitation, then a new account of the description of nature is required in order to achieve a more complete physics.

In 1958 Bohr was able to tie these implications of the quantum postulate together more concisely:

A new epoch in physical science was inaugurated, however, by Planck's discovery of the *elementary quantum of action*, which revealed a feature of

[33] *APHK*, pp. 17, 19. [34] Ibid., pp. 33, 40.

wholeness inherent in atomic processes, going far beyond the ancient idea of the limited divisibility of matter. Indeed, it became clear that the pictorial description of classical physical theories represents an idealization valid only for phenomena in the analysis of which all actions involved are sufficiently large to permit the neglect of the quantum.[35]

Bohr does not explore the implications of the quantum of action ontologically so much as epistemologically. His interest is in the communication of our knowledge of reality. The givenness of the quantum in nature, however, he accepts as a constitutive element of reality. Missing from his argument at this point is a parallel statement about what cannot but be the case in our communication of our knowledge of nature, a transcendental claim about the necessity of employing classical or everyday concepts in any unambiguous descriptions. This 'zeroth' assertion, as I have termed it, will appear later.

At present, then, the following five propositions are seen to make up Bohr's initial moves in his analysis of the implications of the quantum condition:

(i) The quantum of action is a discovery which is universal and elementary.

In other words, the quantum of action is an indisputable fact of nature like the absolute velocity of light or the universal gravitational constant. It is a fundamental constant in nature which applies everywhere.

(ii) The quantum of action denotes a feature of indivisibility in atomic processes.

(iii) Ordinary or classical descriptions are only valid for macroscopic processes, where reference can be unambiguous.

The claims made here have been partly dealt with in section 2.9. above. On the classical scale the effects of the quantum of action can be neglected and a provisional distinction between observer and observed can function adequately. Classical theory can thus be given a 'pictorial' interpretation in which theoretical and observational terms can univocally and unambiguously be applied, but such applications are invalid if a sharp distinction cannot be drawn in the process of observation between observer and observed. And this, Bohr argues, is the case in quantum physics.

[35] *Essays*, p. 2, Bohr's emphasis. For parallel passages see *APHK*, pp. 24, 71 f., 85, 90, 98 f., and *Essays*, pp. 10 f., 18, 24, 91 f.

(iv) Any attempt to define an atomic process more sharply than the quantum allows must entail the impossible, dividing the indivisible.

(v) Because of the limit of indivisibility, a new and more general account of description and definition must be devised.

In other words, because the classical frameworks for description are inadequate at the level of quantum physics (itself a more general theory than classical theory in that it embraces classical theory for high quantum numbers), a more general framework for description is required to replace the idealization of classical pictures. This more general model has to reflect the universal and elementary import of the quantum condition. Note also that by 'definition' Bohr does not simply mean a precise account of the meaning of a term. He is referring to the problem of using well-defined classical terms unambiguously in the new context of quantum physics. And, especially in the context of the Como lecture, 'definition' refers to observation in the context of a space-time description.

Of these five propositions, some are more controversial than others. It is up to the theoretical physicists to prove (i) wrong, and that seems unlikely at present. The status of (ii), (iii), and (iv) is more open to debate, as discussed in sections 2.8. and 2.9. above, but Bohr's argument has proved to be an enduring one, supported as it is by the indeterminacy relations. Even Einstein, who disagreed with Bohr on these issues, was forced reluctantly to agree with the consistency of Bohr's case, saying it was 'logically possible without contradiction'.[36]

The last proposition, (v), is the most crucial. At present it appears unwarranted. It rests on something which Bohr is yet to say, which I have defined as his 'zeroth' consideration of the necessary conditions of the possibility of ordering experience with concepts and conceptual frameworks. Bohr's point will be that concepts operate in conceptual frameworks, and that if a conceptual framework changes radically, as is the case with the introduction of quantum indivisibility, then the concepts can no longer be used in the same way. Until this point is made, however, we will have to leave (v) as it stands and return to an assessment of its validity later.

From this last step, Bohr moves on to face questions about description and definition, and, as he puts it, the conditions for unambiguous communication.

[36] See *APHK*, p. 61, and section 4.4. below.

3.5. (b) *The Conditions for Unambiguous Communication*

How can we describe things which we do not see? More pressing still, what justifies our talk about things which we can in principle never experience directly? Bohr related such questions about the conditions for unambiguous communication to 'the problem of how objectivity may be retained during the growth of experience beyond the events of daily life'.[37] His preoccupation with the possibility of objectivity was also connected with an examination of 'what kind of knowledge can be obtained concerning [atomic] objects'.[38] Stage (v) in Bohr's argument above came at the end of his discussion of the implications of the quantum of action for exact measurement. The next step refers back to (v) and attempts to provide a response to the difficulty which it entails. Thus, after proposing the need for a resolution to the problem of observing and describing quantum processes, Bohr continued his line of thought with a more philosophical point. In 1937 he wrote: 'The revision of the very problem of observation in this field, initiated by Heisenberg . . . led to the disclosure of hitherto disregarded presuppositions for the unambiguous use of even the most elementary concepts on which the description of natural phenomena rests.'[39] The shift of emphasis is significant: Bohr is placing Heisenberg's indeterminacy relations and a newly disclosed condition for the use of concepts on a par. He is moving from physics to philosophy, but the movement is a smooth one.

By 'the observation problem' Bohr refers to the impossibility of knowing if our concepts are being used unambiguously when they are used to describe what cannot be observed, namely the state of the system free of interaction with observing instruments. This problem has to be viewed in the light of the conditions ruling our own use of concepts: when we reach beyond the ordinary limit of experience, Bohr claims, we still have to employ concepts anchored in 'this side' of that limit. In 1948, in the essay on his discussions with Einstein, Bohr again raises the philosophical issues and adopts the same strategy:

[To] clarify the peculiar aspects of the observation problem in this field of experience . . . it is decisive to recognize that, *however far the phenomena transcend the scope of classical physical explanation, the account of all evidence must be expressed in classical terms.* The argument is simply that by the word

[37] *APHK*, p. 67.
[38] Ibid., p. 25.
[39] Ibid., pp. 18 f.

'experiment' we refer to a situation where we can tell others what we have done and what we have learnt and that, therefore, the account of the experimental arrangement and of the results of the observations must be expressed in unambiguous language with suitable application of the terminology of classical physics.[40]

The key word in the last sentence of this quotation is 'can'. It reveals once again how Bohr arrives at his position. He does not argue from premiss to conclusion, but rather articulates his insight into what cannot but be the case through a reflection on what is entailed in his own performance as a communicating agent in, as he assumes, a real world. We 'can' only communicate our experience of quantum events in terms of their interaction with measuring apparatus, because this is the only means by which we are able to observe such events. There is no alternative.

Descriptive terms, Bohr seems to say, must either be related to everyday concepts or to purely theoretical inventions. But a purely theoretical language can have no currency unless it somewhere translates into a set of concepts which have empirical significance. A purely theoretical language is otherwise ambiguous. We are left with only one recourse, it seems, and that is to 'classical concepts', by which Bohr means everyday concepts and their refinements. Bohr is not implying that we are limited to using our present set of concepts, or that we cannot invent new concepts, but he is making the point that any such concepts must be based in some sort of experiential framework if they are to be unambiguous. It is the continuity of macroscopic objects in space and time which makes language stable, he might argue, and which gives basic descriptive concepts their univocity. (As an aside, it is interesting to consider the links between the presuppositions of classical physics—objectivity and continuity of motions in a causal, space-time framework—and these reflections on the basis for successful descriptive language.)

It is in the essay from 1958, in particular, that Bohr stresses the simple logic of his demand:

The decisive point is to recognize that the description of the experimental arrangement and the recording of observations must be given in plain language, suitably refined by the usual physical terminology. This is a simple logical demand, since by the word 'experiment' we can only mean a procedure regarding which we are able to communicate to others what we have done and what we have learnt.[41]

[40] Ibid., p. 39. [41] *Essays*, p. 3.

Bohr is dealing with two separate but related issues in each of these passages. In the light of the conditions implied by the quantum postulate, steps (i)-(v) above, he is aware that the ordinary usage of classical descriptions only applies where reference is unambiguous. He considers that descriptive concepts of this kind, arising as they do in the world of our ordinary experience, are the only terms available to us if objectivity is to be retained in our language. Secondly, given the limitation which the quantum discontinuity imposes, we are only able to speak unambiguously about quantum events when interaction with macroscopic apparatus occurs. What we are then talking about is the whole process of observer/observed interaction, and reference must therefore be made to the conditions of observation if the communication of what is observed is to be unambiguous and accurate.

As a parallel case, let us consider the conditions which would operate if I wanted to communicate unambiguously with the Jindjiparndi people of the inland Pilbara region in north-west Australia. There is no point in my trotting off to Leipzig and speaking in German. Communication requires some sort of sharing of experience and language, whether verbal or signed. When the linguist Carl Georg von Brandenstein came from Leipzig to the Pilbara, he had to learn a new language: 'pändurana', for example, meant 'morning star', and 'marnda' meant 'hill'. We learn words by learning to use them in context. But I do not want to embark upon a Quinean discussion of whether words draw their meaning from the world or vice versa, or simply from their coherence in a language. The realm of the quantum is even further from the realm of classical physics than the Pilbara region is from Prussia. And, whereas the Pilbara and Prussia both belong to human inhabitants, and therefore to the everyday world in which the refined terminology of classical physics can be univocally applied, the quantum world is, as it were, beyond the bounds of immediate human experience: what we learn about the everyday world we learn in a situation where the effect of the quantum condition can be neglected and a distinction between subject and object can be assumed to operate adequately.

Quantum physics entails more than a shift between parallel realms of experience and language, however, because observer and observed are now tied into an indivisible wholeness whenever observation of quantum processes occurs. A different framework therefore applies to the use of our concepts. Once we learn this lesson, we are free to apply these concepts in a different framework. Because Bohr was a realist, because he was not happy with a notion of physics which was purely theoretical, he could never accept the possibility of inventing a new language for quantum physics. Such invented concepts, as opposed to those concepts which can be introduced ostensively, would by definition have no relationship to the everyday world which provides the ordinary context for successful communication and the basis for realism. Whatever concepts were to be used in quantum descriptions, therefore, must belong to the realm of classical or everyday frameworks and their refinements. (On the other hand, it is interesting to note how in recent theoretical physics deliberately enigmatic terms like 'charm', 'beauty', and 'strangeness' have been employed to identify different quarks without in any way suggesting what these quarks might be like.)

Bohr expressed this step in his argument as a 'decisive point' and a 'simple logical demand'. The transcendental character of his philosophy is now made more overt. In a lecture given in 1957 on 'Atoms and Human Knowledge', he states that: 'The main lesson is the simple philosophy, namely that it proved necessary in this new field of knowledge to play closer attention to the conditions which allow an unambiguous use of our physical concepts. . . . [In] almost any field *it is necessary to pay attention first to the conditions which allow a well-defined use of our conclusions*.'[42] The appeal to necessity and to conditions is explicit here. Bohr's concern with conditions, moreover, goes beyond the mere question of experimental conditions. In the draft of his essay on 'Causality and Complementarity', for example, he substitutes the word 'conditions' for both 'arrangement' and 'foundations', thus indicating a sensitivity to both experimental and metaphysical conditions.[43] The problem of observation has thus 'demanded a renewed analysis of the presuppositions for the application of concepts used for orientation in our surroundings'.[44]

A comparison of the quotations above is also helpful for clarifying Bohr's usage of the term 'classical'. I have already shown that Bohr means something quite general: the everyday pictures and concepts of ordinary communication. Refinement by classical physics has not removed them on to a different plane; they have simply been given greater precision. Questions can be asked, of course, about the relationship between these 'visualizables' and the reality they may be taken to

[42] Address given at Macalester College, 11 Dec. 1957. *MSS* : 22, p. 9 of transcript, my emphasis.

[43] Draft of 'Causality and Complementarity', *MSS* : 17, pp. 5 ff.

[44] *APHK*, p. 98. For parallel passages, see *APHK*, pp. 26, 74, 90, 99, and *Essays*, pp. 11, 18, 25, and 93.

represent. Is one thought to be the mirror of the other, or is the difference between image and reality to be viewed more holistically? Bohr's interest, certainly, is focused as much on our use of concepts as on the status of the concepts themselves. Bohr relates the usage of concepts to our participation in the world we experience. This appeal to our practice can be taken to indicate both a holist approach and a transcendental claim, rather than a dualist view of knowledge and mere conceptual explication. More of this will be said in section 3.8. below.

The first stage of Bohr's perennial defence of his interpretation of quantum theory concluded with the 'problem of observation' where sharp separation between observer and observed is impossible:

(v) Because of the limit of indivisibility, a new and more general account of description and definition must be devised.

In the consideration of the conditions for unambiguous communication, Bohr has produced a sixth step, a 'simple logical demand':

(vi) It is a necessary condition for the possibility of unambiguous communication that suitably refined everyday concepts be used, no matter how far the processes concerned transcend the range of ordinary experience.

The term 'simple' once again provides a clue to Bohr's mind. The argument is not an axiomatic one, but a direct articulation of what he sees as constitutive of experiment and communication. There is the same combination of a priori and synthetic judgements which both characterizes and veils transcendental claims. Bohr is certainly open to criticism here, for his argument seems philosophically unsophisticated. He takes it as read that ordinary language can be used to provide a satisfactorily objective account of experience and activity in the real world. Further, he appears to assume that any successful communication must employ terms which are related to such ordinary language. Any relativity in such language is given sufficient success by the reality of nature, *et violà tout*.

It is important to note that at best Bohr has only established a part of his case: it is one thing to make description *unambiguous*; it is a different matter to make it *complete and exhaustive*, and thus achieve closure of his interpretation of quantum theory. Bohr takes up this problem in the next stage of his reflections, and delivers what appears to be nothing more than a pragmatic prescription for the completeness of description.

3.6. (c) *A Prescription for Complete Description*

The last two steps in Bohr's argument, (v) and (vi) above, took him to the edge of a precipice. He made two strong assertions; now he must accept the consequences. Following from (v), he has to devise a new account of description and definition for events beyond the boundary of everyday experience and the realm of classical physics with its causal, spatio-temporal framework. From (vi), he finds himself required to stick with the concepts derived from the everyday world. In section 2.7. above we noted Bohr's apparently pragmatic introduction of the general framework of complementarity. Here we uncover more of the rationale lying behind this 'new account of description'. All the same, at this stage of his argument, the introduction of complementarity appears as a leap out into the unknown.

In 1937, after remarking on the 'observation problem', Bohr continued:

> An immediate consequence of this situation is that observations regarding the behaviour of atomic objects . . . cannot in general be combined in the usual way of classical physics. . . . Far from being inconsistent, the aspects of quantum phenomena revealed by experience obtained under such mutually exclusive conditions must thus be considered complementary in a quite novel way. The viewpoint of 'complementarity' does, indeed, in no way mean an arbitrary renunciation as regards the analysis of atomic phenomena, but is on the contrary the expression of a rational synthesis of the wealth of experience in this field, which exceeds the limits to which the applicability of the concept of causality is naturally confined.[45]

The element of 'indivisibility' renders each observation-event unique and permits the transgression of the usual causal constraints on consistency of description. As discussed above, the limit to the applicability of causality is the point at which no sharp distinction can be made between observer and observed. The cohesion given by the classical framework is no longer possible if the determinist union of causal and space-time accounts is broken. The possibility of a single picture universally covering all aspects of observation has to be abandoned. The manufacturing of a different conceptual scheme is also out of the question, as we have seen, since a purely theoretical scheme would have no basis in reality, and, secondly, any scheme based in reality will be established in the ordinary everyday world, which brings us back to everyday classical concepts and their refinements.

[45] *APHK*, p. 19.

Bohr makes the same case more concisely in the essay from 1948. After discussing the implications of the quantum condition he observes: 'Consequently, evidence obtained under different experimental conditions cannot be comprehended within a single picture, but must be regarded as *complementary* in the sense that only the totality of the phenomena exhausts the possible information about the objects.'[46] The paradoxical nature of this prescription is conceded in the third of the essays under consideration:

> Within the scope of classical physics, all characteristic properties of a given object can in principle be ascertained by a single experimental arrangement, although in practice various arrangements are often convenient for the study of different aspects of the phenomena. . . . In quantum physics, however, evidence about atomic objects obtained by different experimental arrangements exhibits a novel kind of complementary relationship. Indeed, it must be recognized that such evidence which appears contradictory when combination into a single picture is attempted, exhausts all conceivable knowledge about the object.[47]

In each of these passages Bohr flexes the strong arm of rhetoric: this must be done, he asserts, and he uses 'must' in every case.[48] If this entails a rational argument, it is not readily identifiable at this stage of his chain of reasoning. Stapp has suggested that Bohr's approach is best described as 'pragmatic', a view to which Rosenfeld offers no objection.[49] The pragmatism to which Stapp refers, however, is tied especially to Bohr's rejection of both strong objectivism and the mirroring theory of knowledge. The prescription of complementarity does indeed have a pragmatic ring about it, but Bohr is not just washing his hands of the problem of observation. His mind was too scrupulous and too relentless to be satisfied with easy ways out. The prescription of complementarity is Bohr's way of articulating what cannot but be the case. Bohr thought that he could demonstrate the necessity of his prescription.

Indeed, there are traces of an underlying insight in some of Bohr's writings on this point. One aspect of this underlying insight might be called the 'impossibility consideration', something akin to the indispensability claims which lie at the heart of a transcendental argument. In what were possibly the last words he wrote on the introduction of the notion of complementarity, Bohr states that 'the impossibility of

[46] Ibid., p. 40.

[47] *Essays*, p. 4.

[48] For parallel passages, see *APHK*, pp. 26, 74, 90, 99, and *Essays*, pp. 12, 19, 25, and 92.

[49] See Stapp, 'The Copenhagen Interpretation', pp. 1105, 1115.

combining phenomena observed under different experimental arrangements into a single classical picture *implies* that such apparently contradictory phenomena must be regarded as complementary in the sense that, taken together, they exhaust all well-defined knowledge about the atomic object'.[50] This 'impossibility consideration' is not something to which Bohr only appealed late in his thinking. Heisenberg's indeterminacy relations were one stimulus to its formulation, for they formalized the limitation to the applicability of the causal, spatio-temporal framework. In a manuscript from October 1927 Bohr wrote:

It must be remembered, however, that the uncertainty in question is not a simple consequence of a discontinuous change of energy and momentum say during an interaction between radiation and material particles employed in measuring the space-time coordinates of the individuals. According to the above considerations the question is rather that of the *impossibility* of defining rigorously such a change, when the space-time coordination of the individuals is also considered.[51]

Similar formulations are to be found elsewhere in Bohr's writings,[52] but there are also other considerations involved in this 'impossibility' condition. In 1929, reviewing his first attempts at an interpretation of the quantum theory, Bohr omits any physics and appeals to the purely epistemological consideration of the relationship between spectator and actor:

The impossibility of distinguishing in our customary way between physical phenomena and their observation places us, indeed, in a position quite similar to that which is so familiar in psychology where we are continually reminded of the *difficulty of distinguishing between subject and object* . . .

[We] must, indeed, remember that the nature of our consciousness brings about a complementary relationship, *in all domains of knowledge*, between the analysis of a concept and its immediate application.[53]

The prescription of complementarity, therefore, is seen by Bohr to have some connection not only with the circumstances of modern physics, but also with the nature of human consciousness and the problem of

[50] *Essays*, p. 25, my emphasis.

[51] N. Bohr, 'The Quantum Postulate and the Recent Development of Atomic Theory', dated 12 and 13 Oct. 1927, *AHQP* : 36, pp. 4 f. (*NBCW* 6, p. 93), my emphasis. See also the letter to Einstein, 13 Apr. 1927, quoted in section 3.2. above.

[52] See, for example, 'Analysis and Synthesis in Atomic Physics', 25 Feb. 1942, *MSS* : 16, p. 4; Princeton Lecture, 14 Sept. 1946, *MSS* : 17, p. 3; 'Causality and Complementarity', draft, *MSS* : 17, pp. 7 f.

[53] *ATDN*, p. 15, Bohr's emphasis, and p. 20, my emphasis.

distinguishing between subject and object in human knowing. Here we have something quite new being introduced into Bohr's reflections. Indeed, it is the point he takes up next, and in greater detail, thereby showing more clearly the character of his thought.

Before shifting attention to this new development, however, we must consolidate what we have done so far. Bohr has introduced the prescription of complementarity in the following terms:

(vii) Our position as observers in a domain of experience where unambiguous application of concepts depends essentially on conditions of observation demands the use of complementary descriptions if the description is to be exhaustive.

This argument entails several statements. First, it assumes the truth of (vi), namely that the only descriptive concepts we can use are those arising out of the classical world-view. Secondly, it rests on the validity of (ii)-(iv), that each quantum observation is an indivisible whole which is discontinuous with any other observation of the same quantum process, marking a break with the continuous causal space-time framework of classical physics. This is what I have termed Bohr's 'impossibility' condition. Thirdly, responding to the challenge in (v) that a new account of description (or a new conceptual framework) must be devised, Bohr takes the 'impossibility' condition to *imply* that each individual observation is a valid element of a complete description of quantum events. Combining these individual observations is permitted, even where they exhibit strong contradiction, because the classical continuous framework is no longer being applied. Such a mode of combination he calls the complementarity framework.

3.7. (d) *Deep-going Analogy with the Subject-Object Distinction*

Bohr relates the 'impossibility' condition to the 'difficulty of distinguishing between subject and object' and the 'complementary relationship, in all domains of knowledge, between the analysis of a concept and its immediate application'.[54] Indeed, in the course of his various writings on the implications of the quantum postulate, Bohr often refers to the parallel or analogy provided by other fields of enquiry: biology, psychology, and, especially in *ad hominem* debate with Einstein, relativity. These references might appear to be of little significance to the central argument in Bohr's defence of the prescription of complementarity, serving merely as illustrations or

[54] *ATDN*, pp. 15, 20.

examples, but it is clear that from time to time Bohr wants to forge a much closer connection, particularly in the case of psychology.

By 'psychology' Bohr means reflection on the nature of consciousness. In particular, he is referring to the difficulties entailed in subjects making themselves the objects of scrutiny. In so far as the analysis of the problem of subject-object separation provides a parallel to the 'observation problem' in quantum physics, it also lends strength to Bohr's interpretation. Such support, however, is at best indirect. On the other hand, it has already been suggested that Bohr views *both* the discussion of quantum theory *and* the consideration of consciousness as similar instances of a fundamental issue: the problem of communicating objectively about processes which transcend the limits of ordinary experience.

This section will show that, for Bohr, the subject-object problem is not just an *illustration* of the paradoxes introduced by the quantum, but a *similar instance* of the bounds to our experience. In philosophical psychology the subject-object problem is concerned with the paradox of a person describing his or her own consciousness. In quantum physics, as Bohr interpreted its implications, there is the problem of the observer, as a physical system, indivisibly interacting with the observed system. Both sets of conditions are grist for the transcendental mill. The early stages of Bohr's perennial argument, therefore, will now be reconsidered from a more fundamental position.

The elusiveness of the ultimate subject has been a constant source of philosophical speculation. The 'I' as subject can never be turned into an object: immediately one attempts to do so, another 'I' appears to do the searching, and so on *ad infinitum*. This was the point that Bohr took from Poul Martin Møller's story, *The Adventures of a Danish Student*. Some philosophers would dismiss such an approach to the problem as pointless, suggesting that 'I' is simply a reference word like 'here' or 'now', but this was not Bohr's mind.

Reference to the elusiveness of the subject, within the context of the interpretation of quantum theory, appears at least as early as 1927. Bohr concluded his article in *Nature* that year, a revision of the Como address, with the following remarks, pointing to an intimate connection between the two issues: 'I hope, however, that the idea of complementarity is suited to characterize the situation, which bears a deep-going analogy to the general difficulty in the formation of human ideas, inherent in the distinction between subject and object.'[55]

[55] *ATDN*, p. 91; see also pp. 1, 15, 21.

In 1937, in the first of the essays under consideration in this exegesis of Bohr's argument, Bohr couches this analogy in terms of a favourite aphorism of his later years, drawn from Eastern philosophy:

For a parallel to the lesson of atomic theory regarding the limited applicability of such customary idealisations, we must in fact turn to quite other branches of science, such as psychology, or even to that kind of epistemological problems with which already thinkers like Buddha and Lao Tse have been confronted, when trying to harmonize our position as spectators and actors in the great drama of existence.[56]

Ten years later Bohr not only reasserts his earliest views, but also explicitly refers back to his purely philosophical essay of 1929. It was in an article written for the Planck Jubilee in 1929 that Bohr first omitted all physics and concentrated on the purely philosophical. In 1948 he harks back to that essay:

quantum theory presents us with a novel situation in physical science, but attention was called to the very close analogy with the situation as regards analysis and synthesis of experience, which we meet in many other fields of human knowledge and interest. As is well known, many of these difficulties in psychology originate in the different placing of the separation lines between object and subject. . . . A precise formulation of such analogies . . . is perhaps best indicated in a passage [in the Planck Jubilee article] hinting at the mutually exclusive relationship which will always exist between the practical use of any word and any attempts at its strict definition.[57]

There are two separate points being made here. One is related to the comment which Bohr passed to his brother in 1910: 'sensations, like cognitions, must be arranged in planes that cannot be compared'; terms used in one context cannot be uncritically shifted to another context, particularly if the first context is that of everyday objectivity, and the second has to do with the probing of consciousness or of the quantum domain. The second point is that, in the latter context, an absolute separation between subject and object cannot be made.

Thus far, Bohr has simply appealed to analogies, but he takes up the 'hint' in the last sentence quoted himself, and, in the third of the essays under review, he makes the astonishing move of resting everything on the characteristics of conscious individuals.

In general philosophical perspective, it is significant that, as regards analysis and synthesis in other fields of knowledge, we are confronted with situations

[56] *APHK*, pp. 19 f. [57] Ibid., p. 52.

reminding us of the situation in quantum physics. Thus the integrity of living organisms and the characteristics of conscious individuals and human cultures present features of wholeness, the account of which implies a typical complementary mode of description. . . . [We] are not dealing with more or less vague analogies, but with clear examples of logical relations which, in different contexts, are met with in wider fields.[58]

The last sentence is pivotal in coming to understand Bohr's position. Complementarity is required by the feature of wholeness not just in quantum physics, but in any circumstances transcending the ordinary bounds of experiential reference where one attempts to apply descriptive concepts. This is because of a set of 'logical relations' or conditions which are entailed in the formation and application of everyday concepts. When applying such concepts to extraordinary processes which go beyond the everyday, sequential, causal frameworks, these conditions must be heeded. In other words, *both* the case of quantum observation *and* the issue of subject-object separation are instances of a more general consideration of the conditions entailed in the ordering of experience, our formation of concepts, and the range of their valid application. Indeed, in one of the parallel passages, Bohr states this point quite directly: he wishes to 'stress an epistemological argument common to both fields'.[59]

Bohr's focus in giving his interpretation of quantum theory, then, was not just on the issues of measurement and objectivity. These might be described as ontological questions about reality. Once the constitutive role of the nature of human consciousness is drawn into discussion, the questions become epistemic. In other words, questions about reality are seen to be connected with an enquiry into the formation of concepts and the range of their applicability. The element of individuality or wholeness has thus turned problems about the ontological significance of waves and particles into questions about the applicability of the concepts 'wave' and 'particle'. Narrowed in this way, Bohr's focus finally came to rest on a consideration of the conditions involved in our formation of concepts and the nature of human consciousness.

Recognizing the direction in which his reflections had taken him, Bohr inverted the whole process and founded his interpretation on this philosophical substratum: beyond the domain of ordinary everyday usage, the proper application of concepts requires that attention be paid

[58] *Essays*, p. 7.

[59] *APHK*, p. 27. For other passages, see pp. 74; 78 f., 91, 101, and *Essays*, pp. 12 f., 21 f., 28, and 93.

to the whole subject-object situation. These more general considerations can now be seen to lie at the basis of his comments on measurement, correspondence, and complementarity, which are only the surface features of his labours. The substratum has to be uncovered if one is to go beyond what appears obscure, prescriptive, or makeshift in Bohr's work.

In my view, the following uncompromising schema for Bohr's train of thought must be accepted. First, the situation in quantum physics and the situation with respect to the subject-object separation *both entail similar instances of the one argument*. Secondly, this argument rests on the analysis of the conditions for the possibility of unambiguous communication about processes at or beyond the boundaries of ordinary human experience. In defence of the first assertion, reference need only be made to the quotations from 1927, 1937, 1948, and 1958, given at the beginning of this section. There is further evidence, of course, as in the exchange of letters between Bohr and Høffding in 1928,[60] but most notable of all is Bohr's essay for the Planck Jubilee, the one in which he had to omit all physics. Here he reveals the foundations of his interpretation and his insight into the general limits entailed in conceptualization: 'a close connection exists between the failure of our forms of perception, which is founded on the impossibility of a strict separation of phenomena and means of observation, and the general limits of man's capacity to create concepts, which have their roots in our differentiation between subject and object'.[61]

What Bohr means by 'conceptualization' is not entirely clear, and this will be taken up below. The chief point I want to make here is that his appeal to what he calls psychology is not merely an illustration. 'It is not a question of weak parallels,' Bohr once said in a lecture, 'It is only a question of investigating as accurately as we can the conditions for the use of our words.'[62] Before examining his attempts at such an investigation, we can state the two further steps that have been uncovered in Bohr's perennial way of thinking as:

[60] See the letters from Høffding, 11 July 1928, and from Bohr, 1 Aug. 1928, *BSC* : 12. For a German translation of Høffding's barely decipherable letter, see Stolzenburg, *Die Entwicklung des Bohrschen Komplementaritätsgedanken*, p. 248.

[61] *ATDN*, p. 96; see also pp. 1, 15, 20, 21, 91.

[62] See the transcript of Bohr's John Franklin Carlson Lecture, 'The Unity of Knowledge', Jan. 1958, *MSS* : 23, p. 13. By 'words' Bohr also seemed to mean 'concepts' and 'pictures': see parallel passages in *APHK*, p. 2, and in sections 5.2. and 5.3. below.

(viii) There is a similar situation in both quantum physics and psychology with respect to the difficulties of objective description or unambiguous communication.

(ix) There is an underlying epistemological argument which applies to both situations, and which both situations exemplify.

Bohr's writings quoted above contain two important but equally obscure statements which I have not really explored: 'the nature of our consciousness brings about a complementary relationship in all domains of knowledge, between the analysis of a concept and its immediate application'; and, '[a] mutually exclusive relationship . . . will always exist between the practical use of any word and any attempts at its strict definition'.[63] Both sentences appear to be making the same point. I am convinced that the defence of the validity of (viii) and (ix) is related to a grasp of what Bohr is saying in these two statements, both of which have to do with 'the conditions for the use of our words'. Here we arrive at the philosophical substratum to his interpretation.

We can thus tentatively add a further step, which is implicit here, and which will become explicit as a 'zeroth' consideration in the following section. For the moment let us call it:

(x) There are general limits and conditions which apply to our ability to create concepts.

What follows now is an investigation of these limits and conditions and, at last, the uncovering of the much-muddied foundations of Bohr's interpretation of quantum theory.

3.8. (e) *The Conditions for the Ordering of Experience*

In their introduction to *Transcendental Arguments and Science*, Bieri, Horstmann, and Krüger make a basic distinction between conceptual relativism (or thoroughgoing holism) and a more traditional outlook, which, they say, 'tries to revive something like the Kantian notion of "transcendental arguments" which are supposed to refute the sceptic by showing, as against conceptual relativism, that certain conceptual or linguistic frameworks have priority over others, and that the application of certain concepts or linguistic structures is a necessary condition for all talk about "knowledge" and "experience" '.[64] Whereas conceptual

[63] *ATDN*, p. 20, and *APHK*, p. 52.

[64] P. Bieri, R.-P. Horstmann, and L. Krüger (eds.), *Transcendental Arguments and Science* (Dordrecht, Reidel, 1979), p. vii.

relativism amounts to the abandonment of foundational epistemology, the more moderate transcendental approach seeks to establish at least some parameters disclosed as constitutive in human cognitional activity. Bohr, for his part, accepts the actuality of our achievements in unambiguous description of an external reality. By examining the conditions for the possibility of such communication, and in particular the use of concepts and the nature of conceptual frameworks, he has convinced himself of the necessity of employing the framework of complementarity in those areas of experience where strong subject-object distinctions are impossible.

So, Bohr has more in common with the transcendentalists than with the purely pragmatic relativists. He believes there is proper and improper use of concepts; and propriety is guided not merely by convention, but also by the nature of our concept-forming and concept-using activity. He thus describes his undertaking as 'an investigation of the conditions for the proper use of our conceptual means of expression'.[65] In his early essays these reflections are not carried through and often appear towards the end of his chain of reasoning. In his later writings, however, his consideration of the role of conceptual frameworks and the content of these frameworks comes at the beginning of his argument.

In the first of the three essays under consideration here, Bohr comments that, even after the discovery of relativity theory, the situation met with in quantum physics is 'entirely unprecedented in the history of science'. This is because the conceptual structure of classical physics 'rests on the assumption, well adapted to our daily experience of physical phenomena, that it is possible to discriminate between the behaviour of material objects and the question of their observation'. In terms like these Bohr frequently hints at the symbiosis of classical mechanics and the required objectivity for the success of everyday communication. (This will be taken up shortly, and will be discussed in detail in the final section of the concluding chapter.) What Bohr does not state directly here is his underlying conviction that conceptual frameworks can prove to be too narrow, and that an alteration to the conceptual framework will alter the conditions for the unambiguous use of our elementary concepts. Instead, he simply indicates that his 'straightforward solution' to the paradoxes of quantum theory has to do with a clarification of the usage of concepts and may be

65 *APHK*, p. 2.

connected with similar problems in other fields of experience:

> the recognition of an analogy in the purely logical character of the problems which present themselves in so widely separated fields of human interest [i.e., physics and psychology] . . . gives us an incitation to examine whether the straightforward solution of the unexpected paradoxes met with in the application of our simplest concepts to atomic phenomena might not help us to clarify conceptual difficulties in other domains of experience.[66]

The 'purely logical character of the problems' is not apparent in Bohr's essay. In the following decades, however, he refines his approach and gives more attention to the necessity of conceptual frameworks in unambiguous communication.

In the essay written in 1948 Bohr introduces the notions of 'form', 'content', and 'formal frame'. By 'form' he means any word, concept, or image learnt from the everyday world and, perhaps, refined in the discipline of classical science. 'Content' refers to the 'extension' of the form: that which in our experience of nature corresponds to the 'form' or concept. The distinction between content and form is roughly parallel to the distinction between empirical content and theoretical meaning. Thus by 'formal frame' Bohr means the context in which the concept has its given meaning, and he intends both a theoretical and a practical context, without a strong distinction between them. Later he will use the equivalent term, 'conceptual framework', meaning 'the unambiguous logical representation of relations between experiences'.[67] Note that 'form' is not used in either a Platonic or an Aristotelian sense, as far as Bohr is concerned, though he is certainly closer to Aristotle's hylomorphism than to Platonic idealism in his position.

Reflecting on 'inherent limitations in human thinking', Bohr writes: 'The lesson we have hereby received would seem to have brought us a decisive step further in the never-ending struggle for harmony between content and form, and taught us once again that no content can be grasped without a formal frame and that any form,[68] however useful it has hitherto proved, may be found to be too narrow to comprehend new experience.'[69] In the framework of Greek cosmology and mythology, for example, the various gods and the forces of nature were linked together. Within the framework of Aristotelian physics, again, the circular

66 Ibid., p. 20.

67 Ibid., p. 68.

68 While the word 'form' here fits the context and makes sense, note that Bohr uses 'frame' in a similar statement: see the quotation below at n. 72.

69 *APHK*, p. 65.

motions of the heavens could be explained by the fact that heavenly bodies were more perfect and a circle, appropriately, entailed the most sublime geometry. Galileo offered a quite different framework and introduced new concepts, though whether this revolution was the result of his imagination, his telescopic observations, or his temper, may remain open to debate. Newton gave this framework further refinement with his formulation of the notion of gravitational attraction between particle-like masses. And then came Einstein's theory of relativity. In each case, forms and frames are revised in the light of both new experience and new theory. But according to Bohr, up until the formulation of quantum physics, each framework essentially included a sharp separation between form and content, so that 'form' or 'concept' was in a one-to-one relationship with 'content' or 'fact'. Thus the unambiguous application of concepts within these frames excluded the simultaneous usage of mutually exclusive terms. This situation was particularly apparent in classical physics, which linked continuity of motions in a causal space-time framework.

Secondly, note that Bohr is not just referring to the pristine framework of a language here. He is not simply making the holist point that words take their meaning from other words, nor is he merely stating a sceptical version of Heidegger's notion of the hermeneutic circle, that words take their meaning from sentences and sentences take their meaning from words. If he embraces these positions, he also most emphatically wants to include the relationship, as he puts it, between form and content, between concept and its extension. His interest is in the way concepts and frameworks orient us in our surroundings and help us to order our experience. This is a consequence of Bohr's unshakeable realism.

Bohr's logic is more clearly stated in the essay from 1958. He makes his point in the very first sentences:

> The significance of physical science for philosophy does not merely lie in the steady increase of our experience of inanimate matter, but above all in the opportunity of testing the foundation and scope of some of our most elementary concepts. Notwithstanding the refinements of terminology due to accumulation of experimental evidence and developments of theoretical conceptions, all account of physical experience is, of course, ultimately based on common language, adapted to orientation in our surroundings and to tracing relationships between cause and effect.[70]

[70] *Essays*, p. 1.

He then goes on to discuss the contributions of Galileo and Newton and the novel situation arrived at with the discovery of the quantum. This leaves us with the task of unfolding the argument implicit in his 'of course' in the last sentence. What is clear to Bohr, once again, is not always so clear to his audience.

MacKinnon has suggested, I think helpfully, that Strawson's descriptive metaphysics provides many of the arguments that Bohr has left as implicit:

> Strawson concludes that a fundamental condition for the possibility of unambiguous communication is that the conceptual core of any language contains a representation of physical reality as an interrelated set of things with characterizing properties and that these things exist in a common space-time framework. . . .
>
> A description of an object together with its characterizing features does not, in general, suffice for unambiguous reference. It leaves open the possibility of massive reduplication. The only way in which unambiguous indentification of individuals can be secured is through a common space-time framework, which the speaker shares with his hearers. My bodily presence anchors the space-time framework for me and supplies my ultimate reference point for the spatial and temporal location of other objects.[71]

This is certainly in harmony with the key assertions made by Bohr, namely, that unambiguous communication demands that concepts should be drawn from communication about everyday experience, and that the usual framework for such concepts is the causal, space-time framework which offers a continuity of experience and the possibility of re-identification of particulars. Secondly, assuming these conclusions are valid, Bohr then asks how we are to communicate when the causal-spatio-temporal framework no longer applies or when the objectivity of usual experience is impossible to delineate. Indeed, Bohr takes Strawson's argument much further: not only does the success of re-identification in language seem to demand the persistence of some particulars in space and time, but it is also connected with the mechanical view of nature entailed in the strong objectivism of classical physics. This will be taken up again in the discussion of what I have termed Bohr's descriptive metaphysics in Chapter 7.

In an essay entitled 'Unity of Knowledge', written in 1954, Bohr begins with a statement of concern: 'the problem of how objectivity may be retained during the growth of experience beyond the events of

[71] E. MacKinnon, *Scientific Explanation and Atomic Physics* (Chicago, University of Chicago Press, 1982), 361.

daily life'. He then continues, in words which echo the remarks of 1948: 'The main point to realize is that all knowledge presents itself within a conceptual framework adapted to account for previous experience and that any such frame may prove too narrow to comprehend new experiences. . . . When speaking of a conceptual framework, we refer merely to the unambiguous logical representation of relations between experiences.'[72] Bohr is making two distinct points about descriptive knowledge. First, with Aristotle, he is implicitly saying 'when the mind is actively aware of anything it is necessarily aware of it along with an image'; or, similarly, as Kant put it: 'thoughts without content are empty, intuitions without concepts are blind'.[73] Bohr's second point is related to the ordering of such concepts (images, pictures, words, or visualizables). At the boundary of experience, our usual conceptual frameworks lose their purchase, and our concepts may have to be connected in a different manner.

Why is this so? If concepts are to have precise content, then this content is derived from what is experienced unambiguously, usually as a result of the successful workings of language or, in science, through the normative objectivity of macroscopic apparatus. Systematic knowledge entails the ordering of these concepts in theoretical frameworks. The frameworks, however, arise out of our grasp of the manifold of our experience of what is 'other' to us. One has no warrant for exclusively applying such frameworks univocally beyond the bounds of experience, if such bounds exist. Bohr believes that both quantum physics and psychology provide instances of such boundaries beyond which absolute separation of subject and object is impossible. In these circumstances 'experiences cannot be combined in the accustomed manner'.[74]

These final considerations may or may not prove to be good arguments, but they certainly represent Bohr's fundamental position. If, as a result of the quantum condition, the sequential, space-time, and causal framework does not suffice, then a new framework must be provided. The notion of 'complementarity' certainly defies ordinary experience, asking us to hold two mutually exclusive concepts together, shattering our usual framework. Ordinarily, for example, we find it intolerable for two competing accounts to be offered of the same event: one must be wrong, we instinctively declare.

[72] *APHK*, pp. 67 f.

[73] Aristotle, *De Anima*, 432a; Kant, *Critique of Pure Reason*, A 51 / B 75.

[74] *Essays*, p. 19.

We are now in a position to revise (x), as sketched above, into (o), a zeroth condition which belongs both at the beginning and at the end of Bohr's perennial argument and crystallizes his 'epistemological lesson':

(o) All knowledge presents itself within a conceptual framework adapted to account for previous experience, and any such frame may prove too narrow to comprehend new experiences.

In the earlier stages of his argument, at (vi), Bohr stressed the necessity of using only those concepts which the everyday world rendered unambiguous. Now his focus shifts to the frameworks in which these concepts are linked together to order experience and to provide for scientific knowledge. He is not making a distinction here between 'inner' and 'outer' processes, between theory and observation, but attending to the set of presuppositions which govern the use of concepts in our discourse: our theoretical frameworks. An alteration to the formal framework, therefore, also alters the conditions for the unambiguous use of our most elementary concepts.

The notion of complementarity, Bohr also wants to say, can be seen to arise out of the nature of our consciousness of what is 'other' to us, out of the unresolvable tension between content and form, between reality and concept, and between experience and theory. Our representations of reality do not so much involve a privileged mental mirroring of external reality, in which object and subject are absolutely distinct from each other, as a successful compromise of language and activity. It becomes more apparent now that complementarity entails much the same point of view as holism, though not in the thoroughgoing sense, for it arises out of the view that we are, with respect to our use of concepts, suspended in language. Both classical mechanics and language are limited, however, and neither gives the other a universally applicable framework. Yet for Bohr the relationship between word and world is not seen as entirely relative, with the implication that our words have no anchorage in world; instead, given the nature of our consciousness of what is demonstrably 'other' to us, a relationship between word and world is accepted as necessarily defying complete resolution. These considerations, as much as the introduction of the quantum, expose the limits of classical determinism and its assumptions of strong objectivity in every realm of the description of experience.[75]

[75] In these reflections Bohr's work parallels much recent philosophy of science, based particularly on the work of Kuhn, Feyerabend, Toulmin, Duhem, and Quine, which has undermined the key assumptions of scientific explanation, namely naïve realism, the notion of a universal scientific language, and a correspondence theory of truth. See, for

3.9. *Summary*

In this concluding section the various stages of Bohr's consistent presentation of his case will be drawn together and then simplified. Some brief comments on the force of the argument will be made, but the work of evaluating Bohr's position will be left to the final chapter, after the consideration of other aspects of his vision.

The exegesis of Bohr's essays shows that Bohr constantly argues through a series of ten steps, which he expresses in these terms:

(o) All knowledge presents itself within a conceptual framework adapted to account for previous experience, and any such frame may prove too narrow to comprehend new experiences.

(i) The quantum of action is a discovery which is universal and elementary.

(ii) The quantum of action denotes a feature of indivisibility in atomic processes.

(iii) Ordinary or classical descriptions are only valid for macroscopic processes, where reference can be unambiguous.

(iv) Any attempt to define an atomic process more sharply than the quantum allows must entail the impossible, dividing the indivisible.

(v) Because of the limit of indivisibility, a new and more general account of description and definition must be devised.

(vi) It is a necessary condition for the possibility of unambiguous communication that suitably refined everyday concepts be used, no matter how far the processes concerned transcend the range of ordinary experience.

(vii) Our position as observers in a domain of experience where unambiguous application of concepts depends essentially on conditions of observation demands the use of complementary descriptions if the description is to be exhaustive.

(viii) There is a similar situation in both quantum physics and psychology with respect to the difficulties of objective description or unambiguous communication.

(ix) There is an underlying epistemological argument which applies to both situations, and which both situations exemplify. And this is:

example, Hesse's summary of this change in her Introduction to *Revolutions and Reconstructions in the Philosophy of Science* (Brighton, Harvester, 1980), pp. vii ff.

(o) All knowledge presents itself within a conceptual framework adapted to account for previous experience, and any such frame may prove too narrow to comprehend new experiences.

Note that (o), the zeroth condition, both opens and closes Bohr's perennial way of arguing. One can see that (o) is required at the outset if (v) is to have any force. Stage (vi) is another fundamental claim, which allows the transition to (vii). If we cannot change the concepts we employ, then we must change the conceptual frameworks. And the last step, (o), makes even more sense once (vi) has been accepted. The fact that (o) and (vii) are interdependent indicates again that Bohr's argument is not an axiomatic one, proceeding in steps from premiss to conclusion, but a circular one, entailing reflection on what is constitutive in our unambiguous communication.

After recognizing these links, it is possible to arrange Bohr's argument more economically, so that the connections between key statements become more apparent. If the above ten steps reveal how Bohr *did* reason, the following three propositions are intended as a more concise statement of his thought. To begin with, there is the transcendental claim identified as his zeroth condition:

(B1) Some kind of conceptual framework is a necessary condition of the possibility of ordering experience.

The second proposition is a refined version of the Correspondence Principle and of (vi):

(B2) It is a necessary condition of the possibility of objective description of processes at the boundaries of human experience that concepts related to more normal experience be employed.

Finally, the third key statement has to do with the prescription of complementarity, which is now seen to rest on two prior statements. Thus (vii) can now be expressed as:

(B3) Our position as observers in a domain of experience where unambiguous application of concepts depends essentially on the conditions of observation demands the use of complementary descriptions if the description is to be exhaustive.

What is most significant about these propositions is that no mention whatsoever is made of quantum theory. Nor, for that matter, is the separation of subject and object referred to. These cases are simply instances of the more general philosophical consideration of the connection between unambiguous communication and classical mechanics which Bohr began to articulate in 1927. What was of special interest to him, nevertheless, was that the limiting formalism of quantum physics, as created by Heisenberg, pointed to a congruence in two very different areas of human questioning. Puzzles about the description of self-awareness and about the implications of the quantum were suddenly drawn together. This was the 'hitherto disregarded' discovery on which Bohr placed much emphasis. Microphysics and metaphysics revealed the same feature of wholeness or indivisibility which prevented an absolute distinction between subject and object, observer and observed. The formalism of physics was found to support the conclusions of philosophy. And that represented quite a turn-about in the history of the relationship between the two disciplines.

If nature were Newtonian through and through and quantum physics had never been formulated, then the argument for complementarity would be challenged by the universality and continuity of the causal-spatio-temporal framework. But Bohr did not go into such hypothetical considerations, nor does he claim that recourse to complementarity would be inevitable, as a result of the nature of consciousness, whether or not classical theory was universally valid.

Max Jammer, with approval from Heisenberg, summarizes Bohr's way of arguing in only five steps:

1. Indivisibility of the quantum of action (quantum postulate).
2. Discontinuity (or individuality) of elementary processes.
3. Uncontrollability of the interaction between object and instrument.
4. Impossibility of a (strict) spatio-temporal and, at the same time, causal description.
5. Renunciation of the classical mode of description.[76]

Jammer's five steps are almost identical to the stages (i) to (v) outlined above. Without denying the accuracy of his summary, however, I have shown here that a wider context must be given if we are to grasp the full scope of Bohr's thought. The propositions (B1), (B2), and (B3) must be the focus for an evaluation of the validity of Bohr's claims.

In contrast to Weizsäcker, then, I am arguing that Bohr *did* know what he was doing and why it was assertable. Also, *pace* Folse, I think some Kantian aspects of complementarity have been uncovered, especially in (B1). Both Bohr and Kant considered the necessary conditions which apply to the possibility of our using descriptive concepts about the world. And, if Bohr's argument is as purely

[76] Jammer, *The Philosophy of Quantum Mechanics*, p. 101.

philosophical as I am suggesting, then this also implies that any debates about Bohr's interpretation must be philosophical rather than scientific.

In 1958, reflecting on the interplay between philosophy and science, Bohr formulated what amounts to a self-description: 'The task of philosophy', he wrote, 'may be characterized as the development of conceptual means appropriate for communication of human experience.'[77] The idea of complementarity is just such a development, and Bohr arrived at it through a series of transcendental considerations. If one is to identify Bohr in the sweep of philosophies, then it is valid to describe him as a transcendental philosopher.

Before evaluating this philosophy in detail, there are three aspects of Bohr's work which merit special consideration. In the following chapter Bohr's interchanges with Einstein will be reviewed, and attention will be given to his notion of reality as well as to more recent physics which sheds light on the thought-experiments concocted during the Bohr-Einstein debates. After that, in Chapter 5, Bohr's various remarks on the philosophy of science will be gathered together, followed by a chapter in which his ambivalence about the 'mysticism of nature' will be explored. In the light of these considerations, finally, we will turn back to the assessment of his position among philosophers.

[77] Note dated 4 Jan. 1958, *MSS* : 23.

4

Bohr contra Einstein

> To believe this is logically possible without contradiction; but it is so very contrary to my scientific instinct that I cannot forego the search for a more complete conception.
>
> Albert Einstein

4.1. *First Encounters*

The antagonists were the most admired physicists of their time. The prize at stake was nothing less than the certitude of our knowledge of physical reality. For thirty years, with the greatest courtesy, they duelled with one another. The younger man, Bohr, was thought by most observers to get the upper hand; but Einstein regarded these as Pyrrhic victories, since any advance which his adversary made was gained, he thought, at a price which science could less and less afford.

It is wrong, however, to regard Bohr and Einstein as occupying opposing ground on every issue. Though the profundity of their debate over the implications of quantum theory for the possibility of objective knowledge is matched in the annals of physics only by the Clarke-Leibniz correspondence, this does not justify the view that they opposed each other on every question about the nature of physics. The incomprehensible comprehensibility of nature, as Einstein put it, was something upon which both concurred. If Einstein sought for greater comprehension through his defence of classical determinism, Bohr adroitly defended the limits to classical frameworks and offered an alternative means of comprehending our experience of nature. Or, to put it in terms which will be familiar by now, where Einstein searched for sharp delineation between actor and spectator, Bohr continued to assert the mutuality or wholeness entailed in our exploration of nature at the quantum level.

In their debates over quantum theory and reality, Einstein appears, on the surface at least, to be more of a realist and Bohr a positivist. This will become evident in the ensuing discussions. But, as I have demonstrated at length in the previous chapter, there is more to Bohr's philosophy than simple positivism. It will be argued later that, from a more general point of view, both Bohr and Einstein held much the same position in their contemplation of the infinite harmony of nature.

Both men, hagiography aside, were humble dreamers. Neither of them, we pedestrians might judge, had his feet on the ground. Einstein, weary of the adulation he received later in his career, once gave his occupation as artist's model. Bohr, working under the alias of Nicholas Baker on the Manhattan Project at Los Alamos, was given a bodyguard whose occasional employment was to prevent the dreamy 'Uncle Nick' from crossing the road against red lights.

Although they were not often together and corresponded rarely, they displayed a great mutual affection and admiration. Helen Dukas, Einstein's long-serving secretary, remarked that: 'They loved each other warmly and deeply.' Their first meeting occurred in Berlin in the spring of 1920. Shortly afterwards Einstein wrote to Bohr declaring: 'Not often in life has a person, by his mere presence, given me such joy as you'. And Bohr replied: 'To meet you, and talk with you, was one of the greatest experiences I have ever had'.[1] After Einstein's death, Bohr pursued his arguments about the interpretation of quantum theory as though his old adversary was still present and ready to reply.

Their characters were very different in many ways. Einstein was solitary in his work, Bohr gregarious. Einstein sought simplicity, order, and clarity where Bohr, who certainly sought the same order and clarity, seemed to relish paradox and profundity. Einstein showed greater faith in mathematical purity, while Bohr, on the other hand, gave priority to actual physical observations. There are several marvellous photographs of the two of them, taken by their mutual friend Paul Ehrenfest, and in nearly every one their attitudes and expressions contrast with each other: one sits forward and frowns, the other leans back and holds forth; one looks up at the heavens, almost in intercession, the other stares despondently at the carpet; one gestures vigorously, the other had his first glued to his jaw. A telling photograph shows the pair of them walking together out of step, with Einstein slightly ahead, as if trying to flee from Bohr's inexorable dialectic. Bohr has his mouth open, his finger pointing; Einstein's hands are still, his eyes hooded, his mouth fixed in a whimsical, half-unbelieving smile. Bohr is looking up, his hat brim turned up, his shirt collar turned down, his coat on his arm. Einstein is the reverse: eyes down, brim down, collar up, coat on.

[1] See A. Pais, *'Subtle is the Lord . . .': The Science and Life of Albert Einstein* (Oxford University Press, Oxford, 1982), 416 f., and *NBCW* 3, pp. 22 f., 634, on Einstein's relations with Bohr.

The issues which they debated were, in the beginning, diverse. Bohr initially found Einstein's particle- or photon-theory of light difficult to accept, for it ran counter to the experimental evidence that light behaved as waves do. Theory itself, however coherent and mathematically beautiful, was never enough for Bohr. At the same time, Einstein was deeply suspicious of Bohr's apparent compromise of classical physics in his original quantum theory of the atom. For his part, it is clear, Einstein wanted the theory to be more complete in its account of nature.

Then, after Bothe and Geiger had demonstrated experimentally that the Bohr-Kramers-Slater proposal was wrong, and after Compton had reported the results of his scattering experiments, Bohr came to accept the photon-theory of light. This led in turn to the prescription of the framework of complementarity, both with respect to the wave-particle character of light and the correspondence of classical and quantum physics.

Thirdly, in 1927 Einstein began his long campaign to undermine Heisenberg's indeterminacy relations and the mere probabilism which quantum theory seemed to be imposing on physics. At first he attempted to demonstrate the *inconsistency* of the theory. Later, in a paper written with Boris Podolsky and Nathan Rosen, he argued for the *incompleteness* of the quantum formalism.

These three phases of their discussions will be studied in turn in the following sections of this chapter. The final section looks at some of the more recent physical theory which has arisen as a result of the Bohr-Einstein debates over consistency, completeness, and the nature of physical reality.

The titanic scale of these encounters means that they have been the focus of several studies. The most remarkable account, albeit partial and retrospective, comes from Bohr himself in his essay, 'Discussion with Einstein on Epistemological Problems in Atomic Physics', in 1949.[2] My own attention is centred on the underlying philosophical

[2] In P. A. Schilpp (ed.), *Albert Einstein: Philosopher-Scientist* (Library of Living Philosophers 7, Evanston Ill., Northwestern University Press, 1949), 201-41, reprinted in *APHK*, pp. 32-66. See also M. J. Klein, 'The First Phase of the Bohr-Einstein Dialogue', *Historical Studies in the Physical Sciences* 2 (1970), 1-39; M. Jammer, *The Philosophy of Quantum Mechanics* (New York, Wiley, 1974), 109-97; C. A. Hooker, 'The Nature of Quantum-Mechanical Reality', in R. G. Colodny (ed.), *Paradigms and Paradoxes* (Pittsburgh, University of Pittsburgh Series in the Philosophy of Science 5, 1972); E. MacKinnon, *Scientific Explanation and Atomic Physics* (Chicago, University of Chicago Press, 1982), 299-390; K. Stolzenburg, *Die Entwicklung des Bohrschen Komplementaritätsgedanken in den Jahren 1924 bis 1929* (Stuttgart, Ph. D. thesis, 1977), 184-211; and Pais, '*Subtle is the Lord*', pp. 403-59.

positions of the strategies adopted by Bohr and Einstein in their manœuvring through the physics. The focus is not so much on the physics as on the difference between Einstein's ontological defence of realism on the one hand, and Bohr's epistemological defence of realism on the other. It will become apparent that both wished to defend the realism of physics. Their points of view were so different, however, that it was almost impossible to broach the fundamental issues of the discussion.

So, this chapter is chiefly an assessment of Bohr's argument with Einstein in the light of the conclusions arrived at in the preceding chapter. We are concerned here with instances in which Bohr was forced to defend himself on Einstein's grounds rather than from within his own epistemological bastion. How he then shifts the contest back to his own position of strength is particularly illuminating. Bohr turns Einstein's criteria for reality into a discussion of the conditions for the unambiguous application of descriptive concepts. The results of the work undertaken here prepare the way for the ensuing study of Bohr's realist philosophy of science and the evaluation of his descriptive metaphysics.

4.2. *The Early Rounds: Physics and Light*

Einstein's distaste for the original quantum theory of the atom, and especially for Bohr's use of the Correspondence Principle, was evident in his despairing letter to the Borns in 1919: 'One really ought to be ashamed of its success.' If the continuity of physical systems was challenged, so also the comprehensibility of an ordered, causal account came under fire. By leaving so much unexplained, Bohr was also leaving a great deal to chance. To Einstein, the possibility of knowing exactly what was happening in the atomic orbitals appeared to have been too lightly surrendered.

Einstein had definite views about the order of the universe. There was no room for randomness, fundamental discontinuity, or suspension of causality. If physics was not concerned with such a universe, he would, as he once remarked, rather be a cobbler. These views shaped his appreciation of the nature of human knowing and the character of scientific description and explanations. His understanding of the connection between concepts and experience was not entirely opposite to Bohr's, but his demand for continuity and causality certainly exceeded Bohr's more critical approach. Here lay the seeds of their disputes.

We have heard Bohr's constant appeals to the conditions for unambiguous communication in his interpretation of the quantum formalism. His recognition of the predicament in which our language places us is also tantamount to the abandonment of the possibility of making universal, strong distinctions between subject and object, language and fact, theory and observation, and so on. This I have termed Bohr's 'transcendental holism'. It also entails a weakening of the claims to strong objectivity implicit in classical physics and to the universal applicability of the continuous space-time framework that characterizes classical mechanics.

Einstein sought to make much sharper distinctions between thought and experience. In his earlier days, as he confesses in his 'Autobiographical Notes', he was influenced by Mach and Hume and rejected Kant's suggestion that there are a priori connections between thought and experience.[3] Later, according to his own 'epistemological credo', Einstein maintained a distinction between concept and experience:

> on the one side the totality of sense-experiences, and, on the other, the totality of the concepts and propositions which are laid down in the books. . . . The concepts and propositions get 'meaning,' viz., 'content,' only through their connection with sense-experience. The connection of the latter with the former is purely intuitive, not itself of a logical nature. . . . Although the conceptual systems are logically entirely arbitrary, they are bound by the aim to permit the most nearly possible certain (intuitive) and complete co-ordination with the totality of sense-experiences.[4]

Thus physics 'is an attempt conceptually to grasp reality as it is thought independently of its being observed'.[5] The key word here is 'independently'. Bohr, as we have seen, would demur at such a sharp demarcation, even though he also wished to provide a framework which preserved the ideal of the detached observer. The key to the independence of the conceptual grasp of the system from the state of the system itself is, of course, the confidence that the system continues on in space and time whether observation is made of it or not. And Bohr, in the light of the quantum of action, could not accept this for events beyond the ordinary classical realm.

Einstein believed that the universe was comprehensible: 'The eternally incomprehensible thing about the world is its comprehen-

[3] A. Einstein, 'Autobiographical Notes', in Schilpp, *Albert Einstein*, pp. 13, 21, 53.
[4] Ibid., pp. 11 ff.
[5] Ibid., p. 81.

sibility,' he wrote in 1936; 'The fact that it is comprehensible is a wonder.'[6] By 'comprehensibility' he meant the production of some sort of order among sense-impressions, a result of the application of concepts and the establishing of relations between concepts. This order, however, was not imposed upon nature from 'without', as it were, but is rather a property of nature. Hence the wonder of it. With such a fundamental conviction, it was not going to be easy to persuade Einstein to accept the gaping holes which quantum theory left in the foundations of an otherwise continuous and comprehensible world. Bohr had the same convictions about the infinite harmony of nature, but he thought our comprehension of such a harmony depended on the utilization of the broader framework of complementarity. If that meant limiting continuity of the classical mechanical account, then so be it.

Einstein was optimistic about the future of physics: it would continue on a converging course towards the total comprehension of a totally comprehensible reality; although God was subtle, God did not play dice. Curiously, although his opinions about the nature of physics were strong, Einstein also conceded the 'impurity' of the epistemology entailed in his view of the scientific enterprise: the scientist was at one moment a realist, at another an idealist, and at a third a positivist.[7] Bohr, on the other hand, after determining the epistemological conditions operative within physics, insisted on the limitations to the classical ambitions for objectivity and on the limitations to the univocal usage of descriptive concepts. Einstein separated concept and sense-experience, so that science was seen to offer an account of the world's comprehensibility which mirrored reality. Bohr argued, however, that the mutuality of concept and experience limited the possibility of a totally detached, strong objectivity. As noted in Chapter 3, the quantum of action and the indeterminacy relations provided evidence to support his epistemological argument.

In his seminal paper of 1913 Bohr proposed a disjunction between classical physics and quantum physics: 'the dynamical equilibrium of the systems in the stationary states can be discussed by the help of ordinary mechanics, while the passing of the systems directly between stationary states cannot be treated on that basis'.[8] Einstein was naturally appalled at Bohr's legerdemain in the simultaneous application of rival

[6] A. Einstein, *Physik und Realität*, translated by G. Holton in A. P. French (ed.), *Einstein: A Centenary Volume* (London, Heinemann, 1979), 161.

[7] See Einstein's comments in Schilpp, *Albert Einstein*, p. 684.

[8] Bohr, 'On the Constitution of Atoms and Molecules', p. 7.

methods. The comprehensibility of nature must, in his view, be singular rather than plural: it could hardly admit of continuity and discontinuity at the same time. Einstein thus set about the formulation of statistical rules for radiative transitions between stationary states, hoping to achieve the same kind of solution to the quantum problem as classical mechanics had done with its statistical treatments of radioactive decay and thermodynamics.[9]

According to Einstein's view of physics, the observations obtained by instruments had to be explained in terms of the continuity of the underlying, independent, physical system responsible for such results. For Einstein the quantum postulate was provisional rather than inescapable, yet he never succeeded in providing what he considered to be a more complete theory. Bohr, for his part, hammered away at two points. First, the quantum was an inescapable fact of nature. Secondly, the quantum limit made it impossible for us to speak about quantum events except in terms of their interactions with observing apparatus.

In subsequent discussions, these two approaches resulted in alternative views of quantum theory: *either* quantum theory could be interpreted as giving statistical results for an 'ensemble' of systems, as Einstein would suggest; *or*, as Born had first proposed, it provided probabilistic results for individual observations. Implicit in these alternatives, furthermore, are quite differing views of the nature of our knowledge of physical reality. For Einstein the continuity and objectivity of the theoretical account came from the continuity and independence of the physical system; whereas for Bohr the observer was equally responsible for the continuity, and any imagined system at the quantum level could never be observed as independent. These became the prominent issues in the Bohr–Einstein debates about the nature of physics, objectivity, and realism. The issue of determinism versus indeterminism was not the major source of contention, therefore, and only arose indirectly in the context of the debate about continuity.

During the first stage of their encounters the forced marriage of quantum and classical physics was not the only concern. Einstein was on the defensive and Bohr on the attack on another front: for the first two decades of the century there was no direct experimental evidence to support Einstein's photon-theory of light. In fact, commonly observed interference patterns pointed to the opposite conclusion, namely, that light behaved like waves and was a form of radiation. Bohr thus found it

[9] A. Einstein, 'Zur Quantentheorie der strahlung', *Physikalische Zeitschrift* 18 (1917), 127 f.

impossible to accept the photon-theory as anything more than useful hypothesis.

In his address to the Berlin Physical Society in 1920, which was when he met Einstein for the first time, Bohr made his objections courteously and with unusual indirectness.[10] In his later recollection of the subsequent discussion with Einstein in Berlin, however, Bohr stresses the underlying issues of epistemology, the nature of physical reality, and the abandonment of classical determinism:

When I had the great experience of meeting Einstein for the first time during a visit to Berlin in 1920, these fundamental questions [about light radiation] formed the theme of our conversations. The discussions, to which I have often reverted in my thoughts, added to all my admiration for Einstein a deep impression of his detached attitude. Certainly, his favoured use of such picturesque phrases as 'ghost waves (*Gespensterfelder*) guiding the photons' implied no tendency to mysticism, but illuminated rather a profound humour behind his piercing remarks. Yet, a certain difference in attitude and outlook remained, since, with his mastery for co-ordinating apparently contrasting experience without abandoning continuity and causality, Einstein was perhaps more reluctant to renounce such ideals than someone for whom renunciation in this respect appeared to be the only way open to proceed with the immediate task of co-ordinating the multifarious evidence regarding atomic phenomena . . .[11]

The latter description, of course, of 'someone for whom renunciation in this respect appeared to be the only way open to proceed', was undoubtedly Bohr's roundabout way of referring to himself. Bohr was, after all, ready to renounce the universal applicability of classical physics. In 1919, while in great confusion about the direction in which quantum physics was moving, he declared himself 'inclined to take the most radical or rather mystical views imaginable'.[12] And when he had gathered his thoughts together a year or two later, Bohr noted on several occasions that he was introducing fundamental differences to, and making a radical departure from, previous conceptions of physics.[13]

In the early 1920s it was thought that atoms absorbed and emitted light in quanta, but propagated this energy as waves. In April 1921 Bohr was due to address the third Solvay Conference on the topic of

[10] The address is reprinted in *TSAC*, pp. 20 ff., and *NBCW* 3, pp. 242 ff. See especially Bohr's remarks on p. 22.

[11] *APHK*, p. 36.

[12] N. Bohr, undated draft of a letter to Charles Darwin, never sent, about 1919, quoted in Klein, 'The First Phase of the Bohr-Einstein Dialogue', pp. 20 f.

[13] See *TSAC*, pp. 20, 62, 81; *NBCW* 3, p. 242; *NBCW* 4, pp. 264, 283.

'Atoms and Electrons'. Because of poor health, he was unable to attend. His partially completed paper, delivered by Ehrenfest, reveals Bohr at one point weighing up the dilemma posed by the theory of light-quanta: 'on the one hand, such a conception seems to offer the only simple possibility of accounting for the phenomena of photo-electric action, if we adhere to an unrestricted application of the notions of conservation of energy and momentum; on the other hand, it presents apparently unsurmountable difficulties as regards the phenomena of interference of light'.[14]

Even at this time, then, Bohr was ready to accept that there might be conditions governing the application of descriptive concepts and conceptual frameworks. In particular, the universality of the classical causal space-time framework might have to be questioned in order to account for particle and wave descriptions of light.

The Bohr–Kramers–Slater paper of 1924, as discussed above, was an ill-fated attempt to resolve this dilemma. Bohr had endeavoured in that paper to treat energy conservation as statistical and, by so doing, make a break with classical notions of strict conservation of energy. He was forced to acknowledge the incompleteness of the current physical theory, as he had done before on several occasions. After the failure of that attempt, however, Bohr began to consider the possibility of altering conceptions of space-time descriptions. A decade passed, though, before Einstein took up the implications of an 'incomplete' theory. Einstein's initial response to the 1924 paper came, once again, in his correspondence with the Borns: if Bohr was right, then he would rather be a cobbler or an employee of a gaming-house. In another letter, to Ehrenfest, he stated his difficulties quite plainly: 'A final abandonment of strict causality is very hard for me to tolerate.'[15]

Einstein's trepidation was temporarily allayed by the results of the Bothe-Geiger experiments, and Bohr was forced to abandon the notion of 'virtual' oscillators and his suspension of the principle of energy conservation. Again, given the results of Compton's scattering experiments, Bohr had to accept the photon-theory of light not just as an explanatory hypothesis but as a valid description. Rounds one and two to Einstein.

Einstein and Bohr met again at Leiden in December 1925, where Ehrenfest was their host. They discussed a thought-experiment

[14] N. Bohr, 'On the Application of the Quantum Theory to Atomic Problems', *NBCW* 3, p. 374. See Klein, 'The First Phase of the Bohr-Einstein Dialogue', p. 19.

[15] A. Einstein, letter to Ehrenfest, 31 May 1924, quoted in Klein, 'The First Phase of the Bohr-Einstein Dialogue', p. 33.

involving the propagation of light through a dispersing medium, and found that both wave and particle models gave similar results. Entirely new factors were about to come into play, however, with the formulation of the 'new' quantum theory in 1925 and the introduction of matrix mechanics and non-commuting operators.

On 13 April 1927 Bohr wrote to Einstein, enclosing a preprint of Heisenberg's paper on indeterminacy. Bohr also reiterated his belief in the crucial limiting role played by the concepts of classical physics: 'This very circumstance that the limitations of our concepts coincide with the limitations in our possibilities of observation, permits us—as Heisenberg emphasizes—to avoid contradictions.'[16] At the very time of writing this letter, Bohr was also in the process of formulating his notion of complementarity and his interpretation of the new quantum theory. We can see in these remarks the strong connection he perceives to exist between the success of ordinary language, based on re-identification of enduring demonstrable macroscopic particulars, and the key assumptions of classical physics with respect to continuity and objectivity.

The stage was now set for the second phase of the Bohr-Einstein debates. The skirmishing was over. In positing bounds to both our concepts and our observations, Bohr was invading Einstein's sacred ground and his conviction about the possibility of comprehending the world in a neo-classical framework.

4.3. *The Solvay Conferences*

Einstein was not present at Como when Bohr delivered his outline of complementarity in 'The Quantum Postulate and Recent Developments in Atomic Theory', but Bohr gave much the same paper only a month later to the fifth Solvay Conference of physicists, held at Brussels from 24 to 29 October 1927. Einstein was due to attend. It was an event which filled Ehrenfest with nervous anticipation. He wrote to Bohr: 'Naturally, above everything else, I want to understand the asymptotic connection between quantum mechanics and classical mechanics as clearly and from as many angles as possible.'[17] Einstein did indeed attend, though his contribution to the formal discussions was brief. The

[16] Letter to Einstein, 13 Apr. 1927, *BSC* : 10. A fuller quotation from this letter is given in section 3.2 above.

[17] Ehrenfest to Bohr, 24 July 1927, quoted in Stolzenburg *Die Entwicklung des Bohrschen Komplementaritätsgedanken*, pp. 194 f., from the German.

official report of the conference is doubly misleading, however, in that Bohr's purported address is in fact a translation of the version which he prepared for *Nature* in 1928, and, secondly, because Bohr and Einstein, behind the scenes, began to test each other out.

Despite the gaps in the formal historical record, there are several first-hand accounts of these encounters. For example, there is Bohr's own retrospective essay, as well as Heisenberg's recollections in his *Physics and Beyond*. The best contemporary evidence, beyond any shadow of a doubt, is to be found in the letter which Ehrenfest wrote to his students just a few days after the conference had finished. After listing the distinguished participants, Ehrenfest then declared:

BOHR towering completely over everybody. At first not understood at all (Born was also there), then step by step defeating everybody.

Naturally, once again the awful Bohr incantation terminology. Impossible for anybody else to summarize. (Poor Lorentz as interpreter between the British and the French who were absolutely unable to understand each other. Summarizing Bohr. And Bohr responding with polite despair.) (Every night at 1 a.m. Bohr came into my room just to say ONE SINGLE WORD to me, until three a.m.) It was delightful for me to be present during the dialogue between Bohr and Einstein. Like a game of chess. Einstein all the time with new examples. In a certain sense a sort of Perpetuum Mobile of the second kind to break the UNCERTAINTY RELATIONS. Bohr from out of the philosophical smoke clouds constantly searching for tools to crush one example after the other. Einstein like a jack-in-the-box: jumping out fresh every morning. Oh, that was priceless. But I am almost without reservation pro Bohr and contra Einstein. His attitude to Bohr is now exactly like the attitude of the defenders of absolute simultaneity towards him. . . .

In private discussions with Einstein, Bohr has developed other very nice points. The following, e.g.: large, massive, rigid reference systems with imperturbable clocks are particularly suited for the fixation of $x\ y\ z\ t$. But at the same time unable to indicate momentum or energy transfer. This is the way the uncertainty relation shows up in classical mechanics (difficult to notice, but quite unmistakable).

By faintly illuminating a small, freely moving body, one may determine its position rather nicely every hour and evaluate the intermediate velocity and momentum with enormous accuracy. Thereby the uncertainty relation APPEARS to be violated. This is, however, merely a misunderstanding. One has here only EVALUATED the momentum for the intermediate time, not actually measured it. . . . Altogether the notion of a 'conceptual tracking of the particle between the moments of observation' should be rejected just as the notion of a 'tracking of a light corpuscle through the wave field between emission and absorption'.

Bohr says: for the time being we only have at our disposal those words and concepts that yield such a complementary mode of description. But at least we see already that the famous INNER CONTRADICTIONS of quantum theory only arise because we operate with this not yet sufficiently revised language. (I know for sure that this last formulation of mine would drive Bohr to COMPLETE DESPAIR.)[18]

Indeed! I think Bohr would argue that no matter how much we revised our language we would be caught in the same predicament. But, apart from this last sentence, Ehrenfest's contemporary and impartial account, given his friendship with both Bohr and Einstein, is particularly important. Not only does it capture the vitality of the Bohr-Einstein contest, but it also shows Bohr turning arguments based on thought-experiments into arguments about observation and language. While it is certainly true that the debates were about indeterminacy, wave-particle duality, and complementarity, and even though Bohr countered Einstein's arguments from physics with further arguments from physics, Bohr's position at this time was already firmly based on his acceptance of the unavoidable interplay between concept and experience in our descriptions of physical reality. To this end Bohr insisted on providing elaborate drawings of mechanical devices used for observing quantum events, as if to emphasize the connection between descriptive concepts and classical apparatus. Implicit in Bohr's tactics here, also, is his belief in the importance of public identifiability of objects in the shaping of unambiguous descriptive language.

Consider the thought-experiment which Einstein proposed at the fifth Solvay Conference: a beam of electrons travelling, say, from left to right, hits a screen with a single slit in it. The light which passes through the aperture forms a diffraction pattern on a second screen, as drawn by Bohr in Figure 4.1.

Einstein argued that this experiment showed the inability of quantum theory to describe satisfactorily the individual electron events. If A and B are two distinct spots on the screen, and if we know that an individual electron arrives at A, then we also know instantaneously that it does not arrive at B. But this implies some sort of action-at-a-distance effect between two separated locations, contrary to the relativity postulate. Any signal between the two cannot be instantaneous, but must take the time required for the signal to travel such a distance. Before the electron

[18] Ehrenfest to Goudsmit, Uhlenbeck, and Dieke, 3 Nov. 1927, quoted in *NBCW* 6, pp. 37 ff. and 415 ff., and Stolzenburg, *Die Entwicklung des Bohrschen Komplementaritätsgedanken* pp. 197 ff.; see also *AHQP*, microfilm 61.

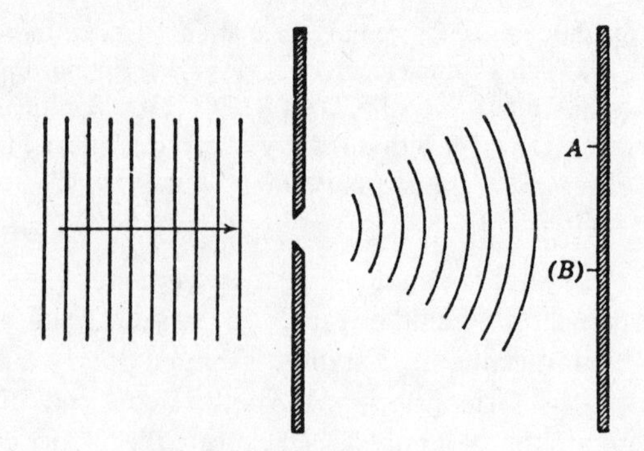

Figure 4.1
Bohr's Version of Einstein's Single-slit Experiment

strikes the screen, it is represented as a spreading, diffracted wave-packet. However, as shown on photographic plates for very low intensities of radiation, the electrons strike the screen as particles might, at specific points, rather than smeared out over a diffraction pattern. This is sometimes referred to as the 'reduction of the wave-packet'.

Quantum theory did not explain why the electron arrived at *A* rather than at *B*. In this sense, it failed to give a complete account of individual processes. Further, quantum theory implied that the final position of the electron on the screen could not be predicted with certainty, but only with probability. Though these predictions could be supported by repeated observations of a 'one-electron' diffraction experiment, Einstein clearly had good grounds for desiring a better theory.

Bohr did not reply directly to this objection, but his later reflections showed the way his mind was moving. When an electron is diffracted at the slit in the first screen, there must also be a change in momentum (Δp), due to interaction between the electron and the screen. Again, the width of the aperture, which denotes the position of the electron and the subsequent wave-cone, introduces an element of uncertainty into the position of the electron (Δq) at the time of the momentum exchange. Appealing to Young's rules for diffraction, as well as to Einstein's derivation of the relationship between the momentum of a particle of light and the number of waves per unit length, Bohr was able to demonstrate the coherence between existing theory and Heisenberg's indeterminacy relations, $\Delta p \Delta q \approx h$.[19] The reduction of the wave-packet

[19] See *APHK*, pp. 43 f.

was thus shown to be consistent with the indeterminacy relations and quantum theory. The only way one could achieve greater certainty in the prediction of the position of impact of the diffracted electron would be either by making a slit of zero width, or by creating an infinite number of diffraction rings (that is, no diffraction at all!). If one accepted the universality of the quantum of action as an element in nature, the theory was as complete as it could be.

In his replies to Einstein, Bohr insisted on presenting pictures of the apparatus which could be used to test the thought-experiments. This was his subtle way of bringing physical theory back to familiar physical reality. Different observations would be made according to the choice of apparatus: a scattering apparatus would indicate particle-like properties, for example, and a diffraction arrangement would give evidence of wave-like behaviour. This is not evidence of inconsistency, however, because the *individual* quantum event is never simultaneously observed as wave-like or particle-like. One cannot perform both a scattering and a diffraction experiment at the same time on the same electron stream. So, Bohr's focus is just as much on the means of observation as on the so-called 'objects' under investigation. He claims that just as we are limited to the options determined by the choice of apparatus, so also we are bound to accept the impossibility of making sharp distinctions between reports of observation and the quantum events themselves.

Einstein, on the other hand, wanted to be able to describe the behaviour of the quantum object itself, independent of its being observed or not. He knew that if he accepted the inevitable 'individuality' or 'wholeness' of quantum events, he would also have to jettison the possibility of establishing a classical statistical treatment of an ensemble of similar events. 'Individuality', as Bohr used the term, implied that each observation was, as it were, disconnected from any underlying system one might like to imagine. Because continuity is threatened by the quantum condition, any statistical determinism is also in jeopardy. For Einstein, if the statistical treatment were lost, then he would also have to surrender that quasi-classical causal account of reality to which his belief constantly drove him. No wonder, then, that he was so reluctant to concede ground to Bohr!

As Bohr later noted: 'Einstein's concern and criticism provided a most valuable incentive for us all to reexamine the various aspects of the situation as regards the description of atomic phenomena.'[20] Shortly

[20] *APHK*, p. 47.

after the fifth Solvay Conference, Bohr set about a more epistemological defence of his position. This was the paper for the Planck Festchrift, in which, as he told Pauli, he left out all the physics and confined himself purely to philosophy. But Bohr also made three references to Einstein's relativity theory in this paper, noting the attention it paid to the observer and the observer's reference system, emphasizing the retention of the classical causal space-time description, and also implying that the introduction of discontinuities into nature was as much Einstein's doing as his own. The quantum theory was the heir to Einstein's action in making the velocity of light a universal 'speed-limit'. It was as though Bohr was deliberately laying bait to draw Einstein back into the contest. Einstein's first response, however, was quite laconic: 'A good joke should not be repeated too often.'[21] Indeed, Bohr's references to Einstein's physics may have obscured the fundamentally epistemological point that Bohr wished to make.

Perhaps Einstein also should have heeded his own advice and not persevered in attempting to undercut Bohr's fundamental argument. At the sixth Solvay Conference, in 1930, he came back at Bohr with another ingenious thought-experiment which was eventually to rebound upon himself. In order to show inconsistency in the indeterminacy relations, he drew upon his formula for the general equivalence of mass-energy,

$$E = mc^2.$$

By measuring changes in mass, it would also be possible to determine changes in energy. If he could simultaneously determine the moment at which this change occurred, then he could disprove the postulated indeterminacy in supposedly non-commuting quantities, energy and time.

Einstein therefore put forward for consideration a box built with perfectly reflecting internal walls, suspended on a balance, and containing a clock. The box was filled with radiation. It also contained a shutter operated by the clock mechanism (Figure 4.2). At a given instant the clock opened the shutter, which released one photon of light-radiation. At the same time, owing to the change in total energy of light within the box, a change in mass would simultaneously be registered on the balance scale. Both measurements could be achieved with arbitrary accuracy. The indeterminacy relations were thus shown to be inconsistent.

[21] See P. Frank, *Einstein: His Life and Times* (New York, Knopf, 1947), 216.

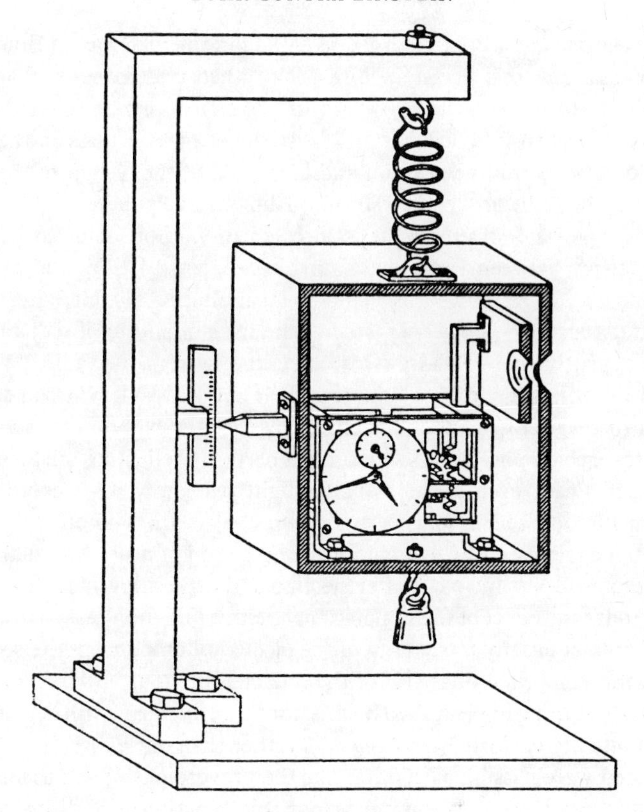

Figure 4.2
Bohr's Version of Einstein's Suspended Radiation Box

Bohr was stunned by this, and spent a sleepless night pondering his next move. Rosenfeld's report brings our dramatis personae to life:

It was quite a shock for Bohr . . . he did not see the solution at once. During the whole evening he was extremely unhappy, going from one to the other and trying to persuade them that it couldn't be true, that it would be the end of physics if Einstein were right; but he couldn't produce any refutation. I shall never forget the vision of the two antagonists leaving the club: Einstein a tall majestic figure, walking quietly, with a somewhat ironical smile, and Bohr trotting near him, very excited.[22]

[22] L. Rosenfeld in *Proceedings of the Fourteenth Solvay Conference* (New York, Interscience, 1968), 232.

The next morning Bohr emerged triumphant: the joke was on Einstein. Bohr realized that Einstein's relativity theory demanded that the relationship between the clock and its position in a gravitational field should be taken into account and, that the change of mass in the box would entail a shift in position and balancing of the weight of the box against the tension in the spring. Allowing for these factors and appealing to a few simple steps in calculation, Bohr showed perfect consistency between Heisenberg's indeterminacy relations and relativity theory.[23] The apparatus could not simultaneously measure, with arbitrary accuracy, both energy and time for a quantum of radiation.

Though Einstein withdrew his objection, he remained unhappy. In 1931 Ehrenfest reported to Bohr that their eminent colleague had added yet a further refinement to his clock-box thought-experiment: suppose that the ejected photon was sent off to a perfectly reflecting mirror some distance away—even as far as half a light-year away. One could now weigh the box over a long period of time, eject the photon, and then weigh it again carefully over another long period of time. Alternatively, one could inspect the clock after ejection and make allowances for time-loss under the effect of the gravitational field. In the first case, one could then predict exactly the energy of the ejected photon, *or*, in the second case, the exact time of arrival of the reflected photon.

Bohr was quite unperturbed by this, for the refinements still demanded two mutually exclusive arrangements rather than one experiment. The two men were passing each other like the proverbial ships in the night: Einstein constantly searching for a possible experiment to disprove the theory; Bohr constantly reiterating the unbreakable wholeness of the experimental arrangement where the effects of the quantum limit came into play. Once again he was back to the conditions for unambiguous communication:

> In fact, we must realize that in the problem in question we are not dealing with a *single* specified experimental arrangement, but are referring to *two* different, mutually exclusive arrangements. . . .
>
> The problem again emphasizes the necessity of considering the *whole* experimental arrangement, the specification of which is imperative for any well-defined application of the quantum-mechanical formalism. . . .
>
> In the quantum-mechanical description our freedom of constructing and handling the experimental arrangement finds its proper expression in the possibility of choosing the classically defined parameters entering in any proper application of the formalism.[24]

[23] See *APHK*, p. 55. [24] *APHK*, p. 57.

A well-defined (and hence useful) descriptive concept depends upon the existence of re-identifiable macroscopic particulars.

Einstein's attempts to show inconsistency in the theory, therefore, ended in failure. As well as meeting Einstein on his own ground, with arguments based in physics, Bohr also tried again and again to explain why quantum physics must be correct. The quantum of action demanded that the indivisible whole of apparatus and quantum system must be taken into account in any description of quantum events. Because of the conditions for ambiguous communication, the only concepts we can use unambiguously are those which arise when a practical distinction can be made between subject and object and where a continuity of reference operates. This is the everyday world out of which classical concepts are shaped. Thus we are stuck with classical concepts and classical apparatus in our descriptions and observations of quantum events. By introducing complementarity, however, Bohr was showing us how to use these concepts in a new way in order to describe the extraordinary world of quantum physics.

Einstein's experimental refinements, nevertheless, set him off on a new train of thought: surely the fact that we could only measure *either* energy *or* time did not mean that the other quantity did not exist. The present theory did not allow us to say anything about this issue. There must be something incomplete about a physics which failed to cope with possibly real values and therefore could not account for possibly underlying realities. (These were later called 'hidden variables' by Einstein's successors, though Einstein himself did not endorse these theories, despite the fact that they would have achieved his goal.) Perhaps there was more to nature here than had yet been discovered.

In 1933 Einstein was forced to leave Germany for America. There he formulated an attack on the completeness of the present quantum theory and prepared his strategy for the final round of his contest with Bohr.

4.4. *Completeness and Reality: The EPR Paradox*

In 1935, in collaboration with Boris Podolsky and Nathan Rosen, Einstein presented a paper entitled 'Can Quantum-Mechanical Description of Physical Reality Be Considered Complete?' In this historic piece of dialectic, later referred to as the EPR paradox, Einstein's intention was no longer to expose quantum theory as untrue, but to demonstrate that it did not tell the whole truth. In this way he hoped to circumvent Bohr and any peremptory closure of the issue. If he could show that there were elements of physical reality upon which the theory did not touch, then, given certain assumptions, Einstein could also assert the *incompleteness* of the theory. And, such a result would contradict Bohr's claims that the theory exhausted all the possibilities for our description of quantum events.

Crucial to Einstein's argument was his *criterion of physical reality*: 'If, without in any way disturbing a system, we can predict with certainty (i.e., with a probability equal to unity) the value of a physical quantity, then there exists an element of physical reality corresponding to this physical quantity.' This criterion is reasonable enough. Typical of Einstein's way of thinking, the argument proceeds from a mathematical theory to a definition of physical reality. Bohr, of course, moved in another direction, for he assumed the existence of physical reality and considered physics to be concerned with ordering our experience of that reality and unveiling its secrets in so far as this was possible. Bohr was later to contest Einstein's condition, 'without in any way disturbing the system', though he did not attack his criterion of reality head-on.

Einstein and his co-authors also propounded a *criterion of the completeness* of physical theory: a physical theory is complete only if 'every element of the physical reality has a counterpart in the physical theory'. If they could demonstrate the existence of elements in 'physical reality' which had no counterpart in the theory, then they would have proved the incompleteness of the theory.

Einstein, Podolsky, and Rosen then proposed a devilish alternative: 'either (1) the quantum mechanical description of reality by the wave function is not complete; or (2) when the operators corresponding to two physical quantities do not commute, the two quantities cannot have simultaneous reality'.[25] Einstein, of course, wanted to defend position (1), whereas Bohr's interpretation appeared to be tied into position (2). If the EPR criterion of physical reality was accepted, however, then the second alternative becomes exceedingly difficult to maintain: a mismatch between the state of the system and the state-function would have been exposed.

The Schrödinger wave-function characterizes the state of a physical system in the quantum domain by using mathematical operators which correspond to observable properties of the system. Einstein's challenge

[25] A. Einstein, B. Podolsky, and N. Rosen, 'Can Quantum-Mechanical Description of Physical Reality Be Considered Complete?', *The Physical Review* 47 (1935), 777-80. This paper and Bohr's reply are reprinted in S. Toulmin (ed.), *Physical Reality* (London, Harper and Row, 1970).

to the completeness of the theory probes the consequences of these non-commuting operators. Consider, for example, the non-commuting operators for position and momentum of an electron. Measuring a definite value for the momentum of the system precludes the possibility of simultaneously measuring its position. Does this imply, Einstein asked, that the electron has no definite position? If so, is it not possible that there are elements in physical reality which the theory cannot account for, and that, therefore, the theory is incomplete?

The EPR paper thus shifted the focus of the debate on to the basic life-blood of physics: the nature of reality and the role of theory. (When in doubt, go for the jugular!) Physicists generally operate with the belief that there is a physical reality independent of their theories, and that their theories are checked by matching them against this physical reality. The physical reality exists, they assume, whether we know it or not. Common sense tells us so, as does the awareness of regularity and continuity in our usual experience of external reality. Whether we attend to it or not, the argument goes, the world does not change if we shut our eyes for a moment. Science, then, is concerned with describing and explaining this distinct external reality. This was, roughly speaking, Einstein's position. It assumes the possibility of always being able to make sharp distinctions between the observed object and the means of observation.

Putting it another way, the question is whether or not physical reality in the quantum domain has the same sort of properties as it is assumed to have at the everyday level: can we, or can we not, ascribe the same sorts of properties to electrons as we do to billiard balls and ballistic missiles?

If we knew of elements in this external reality which science neither describes nor explains, then we would have to agree that the theory was incomplete. This is the point of the completeness criterion. As MacKinnon suggested, a handbook of courses taught at a university would be incomplete if some of the courses which we happened to be attending were not included. Here the theory and the reality can easily be checked against each other. In quantum physics, however, as Bohr laboured to demonstrate, such a sharp separation cannot be made: all that we can observe are the results of experiments which we have designed; as MacKinnon puts it: 'There is no way to sneak a peek at the objectively existing reality and then compare this with what the theory says about it.'[26]

[26] See MacKinnon, *Scientific Explanation and Atomic Physics*, p. 339.

The EPR argument is an attempt to 'sneak a peek' by outwitting the experimental situation and its limitations, as it were. As well as opening up ontological issues, the argument ultimately raises epistemological questions, and, in so doing, it highlights the difference between Bohr's emphasis on the conditions imposed by the nature of our concepts and Einstein's more ambitious hopes for physics. As noted above, and as repeated in the assumptions of the EPR paper, Einstein believed that elements in observed physical reality corresponded to elements of the physical theory: 'every element of the physical reality has a counterpart in the physical theory'. Concepts used in the physical theory thus represented the things in themselves. For Bohr, on the other hand, concepts were related to refinements of language in our unambiguous communication rooted in the everyday world, and such language was itself possible if continuity of reference was provided by both the observer and by nature. There were limits to the application of concepts on a one-to-one basis, and these limits were imposed by the indeterminacy introduced by the quantum of action.

According to the formalism of quantum mechanics, in which physical properties are represented by non-commuting operators, accurate knowledge of one physical quantity can only be predicted by sacrificing the accuracy of the conjugate variable. Einstein and his co-authors, however, proposed an experiment which seemed to enable accurate prediction of *both* quantities. This would then undermine (2) in their devilish alternative, leaving the incompleteness of the theory as a consequence.

The experiment proposed was simple enough. Consider a system of two particles, 1 and 2, which have been in interaction for a given interval of time. The respective momenta and positions of the particles (p_1, q_1) and (p_2, q_2) can be combined in a state with total momentum and relative distance:

$$p = p_1 + p_2$$
$$q = q_1 - q_2.$$

After the particles have interacted, measuring p_1 enables the calculation of p_2 without in any way disturbing the second particle. According to the EPR criterion of reality, this 'certain' figure corresponds to an element of reality. Again, if q_1 is measured, then we can also determine q_2 without in any way disturbing the system. Finally, this certain value of q_2 must equally correspond to an element of reality.

Thus, the EPR experiment seemed to show that both p_2 and q_2 are elements of reality. We are able to discover the position and momentum of the second particle without in any way disturbing it. But this contravenes quantum theory: position and momentum are said to be non-commuting, yet here we have a particle with certain position and certain momentum. Does this mean that the theory is incomplete?

At one level, the EPR argument can be readily dismissed. Bohr's reply was swift and to the point: it is impossible simultaneously to calculate the position and momentum of the unmeasured particle, since it is impossible simultaneously to measure the position and momentum of the first particle. Further, once a measurement is made of the first particle, then the entire measuring system is involved in such a way that, given the indeterminacy relations, it is impossible to claim a meaningful, exact definition of the conjugate property of the second particle.

In the best traditions of debate, Bohr denied Einstein's major premiss. There is an ambiguity, Bohr asserted, in the phrase 'without in any way disturbing a system'. (Note that though Bohr had used the word 'disturb' in his earlier writings, he had abandoned it by this time, since it implied that the quantum system was in a knowable state that could be disturbed.) Einstein, Podolsky, and Rosen attempted to deal with this objection by making what is tantamount to a *locality assumption*: if 'at the time of measurement . . . two systems no longer interact, no real change can take place in the second system in consequence of anything that may be done to the first system'.[27] (More will be said about 'locality' in the discussion of Bell's theorem in the following section.) Measurement of particle 1, so this tacit locality assumption implied, could not instantaneously disturb particle 2, which by now was some distance away.

If we were talking about ordinary physical events, such action-at-a-distance would indeed seem intolerable. But, as Bohr insisted all along, we are dealing with the classical measurement of a quantum event. Quantum theory does not describe the state of the quantum system so much as the state of a whole combined system of quantum 'object' in interaction with classical measuring apparatus. In making what might be called a quantum measurement, one is in fact turning to concepts which have empirical reference through the application of a macroscopic space-time framework. Further, we must concede that we are interfering with the whole quantum-classical system and not just with one part of it. If we choose to measure the momentum of one particle, then we are discontinuously freezing the whole system, since an

[27] Jammer identifies this point in *The Philosophy of Quantum Mechanics*, p. 185.

unanalysable transfer of momentum occurs between the quantum system and the classical apparatus.

The consequences for the second particle, it might almost be said, are not mechanical but conceptual! That is, we have fixed the conditions for unambiguous communication about the second particle by employing a particular observational arrangement on the first. It is true that measurement of a property of particle 1 does not cause a mechanical disturbance to particle 2; but measurement of the momentum of particle 1 means an incalculable alteration to the whole framework of reference of the measuring apparatus, so that an exact measurement of the position of particle 2 with respect to that same framework is out of the question.

The concept of exact position, after all, has its context in a rigid spatial framework. When the rigid framework is disturbed by the measurement of momentum of the first particle, then the possibility of applying the concept of exact position to the second particle is lost. The system has been 'disturbed' because the frame of reference has been 'disturbed'. Bohr summarized his response thus:

> From our point of view we now see that the wording of the above-mentioned criterion of physical reality . . . contains an ambiguity as regards the meaning of the expression 'without in any way disturbing a system.' Of course there is in a case like that just considered no question of a mechanical disturbance of the system under investigation during the last critical stage of the measuring procedure. But even at this stage there is essentially the question of *an influence on the very conditions which define the possible types of predictions regarding the future behaviour of the system.* Since these conditions constitute an inherent element of the description of any phenomenon to which the term 'physical reality' can be properly attached, we see that the argumentation of the mentioned authors does not justify their conclusion that quantum-mechanical description is essentially incomplete.[28]

In other words, if we measure momentum for one particle, we influence the conditions for describing the other particle; we have interfered with the system to such an extent that it is meaningless to talk about the exact position of the second particle. The entire system is, in this sense, linked together; the 'locality assumption' is untenable.

In one move Bohr not only poses an objection to Einstein, but also shifts the ground of the debate. Einstein's notion of locality and reality has to do with, if you like, *physical* isolation. Bohr's notion, on the other hand, is also connected with the possibilities of *conceptual* delineation.

[28] N. Bohr, 'Can Quantum-Mechanical Description of Physical Reality Be Considered Complete?', *The Physical Review* 48 (1935), 696-702; see *APHK*, pp. 60 f.

Classical notions like 'position' are derived from our usual experience of independent regularities in the everyday world.

In his reply to the EPR paper, Bohr moved from defence to attack. The EPR experiment displayed the same limitations as previous thought-experiments: it entailed mutually exclusive arrangements of apparatus. If one was measuring momentum of the first particle, then one could not measure its position, and vice versa. And, while a measurement of the momentum of the first particle allowed a calculation of the momentum of the second particle, such an experiment precluded the possibility of defining the position of the second particle because of the uncertainty introduced into the entire measuring framework. Two separate, *successive* experiments, one measuring momentum and the other measuring position, hardly established the *simultaneous* momentum and position of the unobserved paired particle.

Bohr argued that acceptance of the quantum discontinuity necessitated the imposition of limits upon the applicability of our usual conceptual frameworks: 'the *finite interaction between object and measuring agencies* conditioned by the very existence of the quantum of action entails . . . the necessity of a final renunciation of the classical ideal of causality and a radical revision of our attitude towards the problem of physical reality'.[29] Here Bohr's revolutionary spirit surges forth. The status of physical theory which described reality as a continuously regulated flux of independent sets of objects was now, unequivocably, under threat. Yet Einstein was equally obdurate in his defence of this last bastion of classical physics. He totally rejected the incipient subjectivism which he thought Bohr was introducing into physics. Towards the conclusion of the EPR paper the authors remarked that quantum theory seemed to imply that the reality of p_2 and q_2 depended on the sort of measurement that was made of the first particle: 'No reasonable definition of reality', they concluded, 'could be expected to permit this.' Bohr, for his part, rejected such charges and sought to explain why his position maintained the objectivity of physics, albeit a new kind of objectivity.

Bohr thought that Einstein was seeing the problem from the wrong perspective. In Bohr's mind, the reality was not created by the measurement, nor did what happened in one part of a paired system magically transform what happened at the other. The quantum system could be described only in terms of interaction with classical apparatus. One could not 'sneak a peek' behind the curtain, for there was nowhere

[29] Ibid.; see *APHK*, pp. 59 f.

behind the curtain that one could go. One could not distinguish sharply enough between object and measuring apparatus at the quantum level.

Two separate strands of thought are intertwined in Bohr's response to Einstein. One has to do with the inability of quantum mechanics to control the effects of measurement on the system being measured. (Here the analogue is classical measurement of an independent system, where such interactions are controllable and uncertainty relations have not been heard of.) The second strand of thought, however, includes the idea that quantum mechanical 'objects' should not be thought of in the same way as classical objects, that is, as independent bodies. Because they do not exist in determinate states, such states can never be defined for them: such definition would require mutually exclusive sets of observations. While Bohr uses the first strand of thought, it is the second that is the more decisive thread in his argument.

Although subsequent refinements of the EPR paradox raised further issues, as we shall see, these two publications in 1935 marked the end of the Bohr-Einstein debates. In 1936 Einstein acknowledged that it was 'logically possible without contradiction' to hold that quantum theory offered a valid description of individual phenomena, and in so doing, he extended an olive branch to Bohr. Einstein's spirit remained indomitable, though, for he then added: 'but it is so very contrary to my scientific spirit that I cannot forego the search for a more complete conception'.[30]

Einstein's unhappiness with the complementarity interpretation of quantum theory had several causes. First and foremost, he thought that Bohr was disfranchising physics by denying it the right to describe physical reality as it exists independent of observation. He took this to imply that Bohr had sold out on realism. Secondly, it must have seemed to Einstein that complementarity itself could not be accepted as a principle in physics since its consequences could not be tested empirically. In fact, Bohr was defending a different understanding of the realism of physics, one which was forced on him by the consequences of the quantum of action as a fundamental element in nature. Bohr never disputed the reality of the quantum system described by the wave-function. What he did resist, however, was the unwarranted application of classical concepts to such systems independent of their interaction with measuring apparatus. Otherwise we would have to describe

[30] A. Einstein, 'Physics and Reality', *Journal of the Franklin Institute* 221 (1936), 349. See *APHK*, p. 61.

the electron as both a wave and a particle at the same time, and this would be a nonsense.

Bohr had not, of course, disproved Einstein's view of physics; he had merely defended his own position against Einstein's attack. And, furthermore, as long as the quantum discontinuity remained a fundamental element of physics, Bohr's position would be difficult to overthrow. This last phase of the Bohr-Einstein debates is particularly difficult to evaluate because it does not really touch on the basic differences of outlook between the two men. Discussing their arguments is almost like adjudicating a debate conducted in two different languages, neither of which is understood by the other speaker. All that Einstein's objections demonstrate in the end is the fact that Bohr's position rests on the givenness of the quantum, on the impossibility of strong distinctions between observer and observed, and on the conditions for the unambiguous application of descriptive concepts in such a situation. And for Einstein, on the other hand, the quantum postulate always remained more of a heuristic device than a universal fact of nature.

At a more general level of the debate, however Einstein's strong opinion was sufficient to cast doubt on the completeness of quantum theory. He had placed the key issue on centre stage: if quantum mechanics was a complete theory, then our usual notions of objective reality would have to go. His persistence influenced the proposal of 'hidden variable' theories and the refinement of further experiments by which physicists might test the conclusions of quantum theory. This remarkable turn of events will be discussed in the following, final section of this chapter.

Though the differences between Bohr and Einstein remained unreconciled, these friendliest of rivals were henceforth confined to replaying their old battles rather than being able to open new fronts—except perhaps for one occasion on which Einstein invaded Bohr's territory (which, curiously, was also Einstein's home ground). Bohr was standing at the window of Einstein's office, dictating to Abraham Pais; Einstein chose to work in the smaller assistant's office next door. Bohr was standing at the window and repeating 'Einstein . . . Einstein . . .' as he searched for the words to express his thoughts. Pais takes up the story:

> At that moment the door opened very softly and Einstein tiptoed in.
>
> He beckoned to me with a finger on his lips to be very quiet, his urchin smile on his face. . . . Einstein was not allowed by his doctor to buy any tobacco. However, the doctor had not forbidden him to steal tobacco, and this was precisely what he set out to do now. . . .

Then Bohr, with a firm 'Einstein' turned around. There they were, face to face, as if Bohr had summoned him forth. It is an understatement to say that Bohr was speechless. . . . A moment later the spell was broken when Einstein explained his mission and soon we were all bursting with laughter.[31]

4.5. *Einstein Locality and Quantum Realism*

The world, as we perceive it, is relatively stable: there is a continuity in our perceptions; the tree outside my window is always there when I open the curtain each morning. A person who holds that this regularity in observation is due to some abiding physical reality, whose existence is independent of whether it is observed or not, is a 'realist'. 'Anti-realists', on the other hand, maintain a range of sceptical positions about the realist claims: for example, they may see no justification for stating that things exist which cannot be observed, or they may regard reality as a dream-state and a projection of the mind, or they may believe in the existence of an external world but be sceptical about truth claims regarding the nature of the reality behind experienced phenomena. Those among the positivists who hold the verifiability criterion of meaning, if not denying the existence of an external reality, claim that statements about that reality which cannot be directly related to observation are meaningless. In this sense, these positivists argue, our knowledge of reality is in practice restricted to the content and circumstances of our observations and experiences.

Bohr's response to the EPR paradox seems to have a whiff of positivism about it. Does he not declare, after all, that it is meaningless to talk about the position and momentum of the second particle in the EPR experiment except in the context of an observation on the first particle? The positivist principle, however, is primarily concerned with meaning and the validity of reports of sense-experience. This is not precisely Bohr's interest. Two major issues to which he devoted his thinking were the discontinuity in nature introduced by the quantum of action, and, secondly, the impossibility of making absolutely sharp distinctions between subject and object in human knowing. Both of these, however, were connected with his exploration of the conditions for the unambiguous use of descriptive concepts.

The terms 'realist' and 'positivist' have their context in the discussion of the common-sense experience of an external reality. Classical physics

[31] A. Pais, 'Reminiscences from the Post-war Years', in S. Rozental, (ed.), *Niels Bohr: His Life and Work* (Amsterdam, North-Holland, 1967), 225 f.

sometimes appears to entail a naïve realism, accepting that things are as they appear to be. But as physics became more complicated a more critical approach was required. Positivism was one such response, inviting physicists to speak about what was observed without being troubled by the appearance-reality distinction. In so far as Bohr adopted a critical stance towards the difference between observation and reality, then it might be appropriate to label his position, in some aspects, as 'positivist'. But Bohr is also struggling to establish a different notion of reality. To accuse him of positivism would, given this difference of interest, be rather like a devout cricketer accusing a baseball pitcher of throwing: the accusation may be valid, but what is 'not done' on the cricket pitch at Lord's remains essential to the activities from the mound at Wrigley Field.

In the course of the EPR argument, as we have seen, the authors postulated a criterion of reality and a tacit assumption about locality. The reality criterion expressed Einstein's belief that physical theory ought to parallel the way things are in nature: if we know the value of a physical quantity with certainty, then there also exists an element of physical reality corresponding to this quantity. Implicit in this criterion is a correspondence theory of truth and an abstraction theory of knowledge: the quintessential mind is seen as a mirror of the material world of our sense-experience.[32] Einstein's criterion of reality thus rests on a particular view of physics and a particular view of human knowing. Bohr, on the other hand, struggled to articulate less clear-cut views on both questions and to shape a totally new view of reality.

The 'naïve realist' is one who believes that things are as they appear to be: the spoon in the glass of water actually is bent, the earth is actually flat, and the sun actually moves around the earth. The 'critical realist' asks further questions before judging how things might really be. It would be churlish to accuse Einstein of naïve realism and to identify Bohr with classical critical realism, but the difference between them has its parallels in the differences between naïve and critical realism. (A more detailed discussion of Bohr's 'relative' realism is given in section 5.2. below.)

Einstein's tacit 'locality assumption' came to present a threat to the structure of his own world-view and, in the end, became the Trojan horse for Bohr's campaign. The locality assumption, it will be recalled, stated that when two systems are separated by a distance, a change that

32 For evidence of Einstein's 'abstraction' theory of knowledge, see his letter to Maurice Solovine, 7 May 1952, reprinted in French, *Einstein*, pp. 271 ff.

takes place in one system cannot cause a simultaneous real change in the second. This assumption is related to Einstein's relativity theory and to the impossibility of sending a signal from one place to another at more than the speed of light. The assumption is also expressed in the dictum that nature is 'local': what happens in one locality cannot instantaneously alter the state of affairs in another locality.

In this section I want to review some of the developments which have occurred in physics as a consequence of the Bohr–Einstein debates. While the results of these developments may not yet be regarded as absolutely conclusive, the indications are that we may *either* have to abandon a realist view of the world *or* accept the possibility of some kind of non-locality *in our account* of nature. In other words, Einstein's 'locality assumption' either requires modification or we need to articulate a new notion of 'quantum reality'. Bohr's 'higher order positivism', according to one view of reality, may turn out to be a new kind of realism in a more general account of the physical world. I would also like to point out that the comments here should be regarded as introductory and that many subleties of the physics have, perhaps incautiously, been left aside.[33]

First, let me make a few remarks about 'hidden variables'. Suppose that we observed regular behaviour in some series of events or other. We would suspect that there was an inherent factor at work which caused such law-like activity, and, after doing some research on the matter, we might uncover the key factors responsible for the regularities observed. For example, the regularity in the movements of the planets and the stars was eventually explained with the help of discoveries made by Kepler and Newton. In quantum mechanics, if there are further sub-quantum variables at work which would explain otherwise 'probable' events, the quantum limit itself seems to restrict us from access to them. These factors thus become, by definition, 'hidden' variables.

If quantum theory is complete, of course, then it is inconsistent to hold that there are any hidden variables to be uncovered. Einstein's challenge to the completeness of quantum mechanics, however, gave rise to a variety of speculations about the possibility of hidden variables (though he himself never used the term). The discovery of these variables would offer the possibility of causal, determinist explanation:

[33] Some of these issues are covered in greater detail by A. Shimony in his paper, 'Physical and Philosophical Issues in the Bohr-Einstein Debate', to be published in the Proceedings of the Bohr Centenary Symposium of the American Academy of Arts and Sciences (November 1985). I am grateful to Professor Shimony for sending me a preprint of this paper.

the theory would not be limited to predicting the probable results of observations, but could also explain why one event happened rather than another. For example, in the single-slit thought-experiment, the revised hidden-variable theory could tell us why a diffracted electron arrived at A rather than at B. Determinism would be restored to science.

Recent developments in physics seem to have dashed these hopes. J. S. Bell, using much the same starting-point as the EPR argument, has demonstrated that 'no deterministic local hidden-variable theory can reproduce all of the statistical predictions by quantum mechanics'. Further work also revealed that even non-deterministic theories, under general constraints of locality, were equally incompatible with quantum mechanics. Secondly, as a result of Bell's work and recent experimentation, 'it can now be asserted with reasonable confidence that either the thesis of realism or that of locality must be abandoned'.[34] If this is right, then either way Bohr is vindicated: 'Bohr's position remains as one of the few feasible options concerning the foundations of quantum mechanics.'[35] Moreover, the new view of reality which Bohr struggles to articulate is now seen to be partly based on physics and should not be judged solely on the philosophical arguments normally applied in debates about realism.

Let us now move carefully to these questions. Perhaps the best place to begin is with an account of the kind of experiment required to test the predictions of Bell's inequalities against the predictions of quantum mechanics. It may then be possible to understand a little of the theoretical considerations and, finally, to comment on the notions of reality, determinism, and locality.

In 1971 Bell proposed the following experiment, illustrated in Figure 4.3. Imagine a source that emits paired particles (for example, twin protons so that if one proton is spinning 'up' then the other is spinning 'down'). The pairs are then split, so that each component moves off in an opposite direction to the other. A detecting device, called an 'event-ready detector', monitors each stream of particles and counts them as they pass through its gate. It also sets in process a series of subsequent

[34] These two quotations are from the introduction to the magisterial review paper by J. F. Clauser and A. Shimony, 'Bell's Theorem: Experimental Tests and Implications', *Reports on Progress in Physics* 41 (1978), 1883. See also J. S. Bell, 'On the Einstein Podolsky Rosen Paradox', *Physics* 1 (1964), 195-200. For a more popular account of these issues, see B. d'Espagnat, 'The Quantum Theory and Reality', *Scientific American*, Nov. 1979, 128-40.

[35] Clauser and Shimony, 'Bell's Theorem', p. 1922.

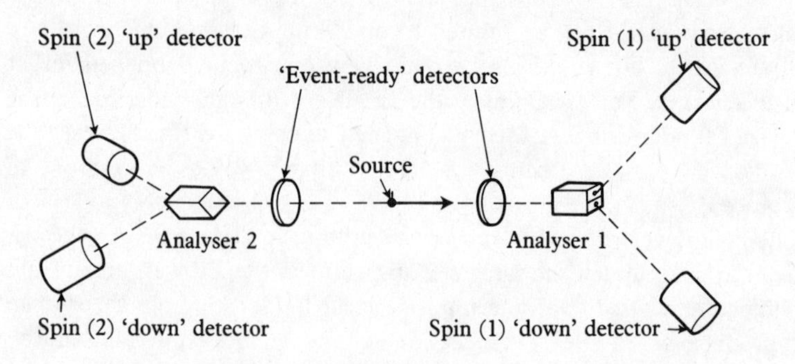

Figure 4.3
Apparatus Configuration (simplified) for Bell's 1971 Proof.

measurements: after passing through the gate of the event-ready detector, the particle enters a spin analyser which then shoots it off either into a 'spin-up' detector or a 'spin-down' detector.

If the particles entering the right-hand system are designated *a*, and the particles in the left-hand system as *b*, then the numbers of *a* and *b* particles, with their registered spins (as well as those which register no spin), can be compared. In an ideal situation, for a given pair, we would be inclined to say that if one particle registers 'spin-up' along a particular axis, then the other will register 'spin-down' along that axis. This might tempt us to claim that we can predict with certainty the result of a measurement at one side of the apparatus if we know what is happening on the other side. By implication, then, we are led to accept realism: my measurement of 'spin-up' on particle *a* can hardly throw particle *b* into a certain 'spin-down' state, for that would contravene locality. Therefore particle *b* must always have been in that state, and must have an independent reality. The EPR spirit lives on!

Bell set about calculating a relationship between the results obtained on the two separate sets of detectors. By including the assumption of locality, he was able to combine information from one set of detectors with information from the other set, and vice versa, and establish an inequality-relationship between the results from different arrangements of the detectors. Quantum theory, however, predicted different expectation values.

All that remained to be done, as a result of Bell's theoretical work, was to set up an experiment and to discover whether the results supported quantum mechanics or not. Such experiments are extremely difficult to arrange, but the handful which have been completed, though not without loopholes, do appear to be very much in support of quantum theory.[36] The implication, then, is that something must be amiss with Bell's premisses (which is not a criticism of Bell, for his argument was designed in order to test the premisses of the EPR paradox).

What were Bell's initial assumptions? The general locality condition is perhaps the easier to understand: if two systems are well separated when measurements are performed, then the value of a measurement on one system cannot be causally affected by what we choose to measure on the other. Secondly, and partly entailed in this condition, there is a realist assumption: it is permissible to speak of a system having a property independent of any measurement of that property. The inclusion of determinist or non-determinist hidden-variable effects, it was found, had no significant influence on the results.

The key premisses, therefore, are locality and realism, the same premisses which were operative in the EPR paradox. If the experiment vindicates quantum theory, then *either* locality *or* realism, as defined here, may be untenable. It does not necessarily follow, however, that *both* must be discarded. As Clauser and Shimony conclude their survey:

> Because of the evidence in favour of quantum mechanics from the experiments based upon Bell's theorem, we are forced either to abandon the strong version of EPR's criterion of reality—which is tantamount to abandoning a realistic view of the physical world (perhaps an unheard tree falling in the forest makes no sound after all)—or else to accept some kind of action-at-a-distance. Either option is radical, and a comprehensive study of their philosophical consequences remains to be made.[37]

Can we find some alternative and helpful proposals in Bohr's work? I think so. For example, in his emphasis on the 'wholeness' of any observation of quantum events, Bohr both anticipated and redefined the issues raised by assumptions about locality. After all, the problem of locality arises at the quantum level only when we make sharp space-time distinctions about the quantum system independent of its interaction with measuring apparatus. To talk of such an independent system, Bohr says over and over again, is impossible, not because realism is denied us, but because the only descriptive concepts we have are those which arise out of the everyday world. The issue, once again, rests on the conditions for unambiguous communication. Secondly, we will not find an abandonment of realism in Bohr's writings, but a

[36] See Clauser and Shimony, 'Bell's Theorem', pp. 1903-18, and d'Espagnat, 'Quantum Theory and Reality', pp. 136 f.

[37] Clauser and Shimony, 'Bell's Theorem', p. 1921.

realism of a different kind. In this context it is important to recall that Bohr's talk of 'quantum phenomena' is not a move away from realism into idealism. It is, rather, a term which describes what can be observed in the real interaction of a real quantum system with a real macroscopic measuring system. What we cannot legitimately talk about, on the other hand, are the 'real properties' of the quantum system independent of its being observed. Such a move would lead us into the trap of imagining quantum 'objects' in the same way that we think of independent macroscopic objects.

Einstein's criterion of reality is a quite specific one. It was part and parcel of his view that physical theory should give an account of physical reality as it exists independent of our observations. Alternative formulations of realism in physics, which Bohr worked towards, are not necessarily, therefore, anti-realist. Einstein suggests that the existence of definite but unmeasured properties implies the existence of independent though unobserved entities. Bell's work, on the other hand, has led us to believe that we would be mistaken in attributing unmeasured properties to sub-quantum entities, thereby lending support to Bohr's position.

In contrast with Einstein's ontological approach to realism, Bohr engages in something like conceptual analysis. He finds himself with little alternative, since, unlike Einstein, he accepts the quantum discontinuity as an inevitable component of physics. He does not begin with an 'inner-outer' distinction and then try to justify the independent existence of that which is 'without'. Instead, he first of all presumes the existence of a reality of which we are a part, and then seeks to establish valid ways of communicating about that reality. Einstein is wrong, according to Bohr, to talk about (say) the position of an unobserved particle on the quantum scale: 'position' is a classically defined property and can only be used unambiguously in the context of classical observation. Thus Einstein is not so much mistaken about his criterion of reality as about his understanding of the valid application of concepts. His quest for an enduring classical 'realism' is inconsistent with quantum theory as it stands, because he projects into the quantum realm concepts which have their meaning rooted in our participation in the everyday world.

Shimony has recently argued that Einstein's position can be broadened to meet the results of the experiments inspired by Bell's theorem: while quantum mechanics is a non-local theory, 'its nonlocality "peacefully coexists" with the relativist prohibition of superluminal signals' if one introduces a modality of potentiality in reality which is intermediate between bare logical possibility and full actuality.[38] Bohr's approach, on the other hand, remains much more epistemological than ontological, more concerned with conditions for unambiguous communication than with the existence of physical things. How such an approach allows Bohr to claim both objectivity and realism for physics is the focus of discussion in the following chapter.

[38] Shimony, 'Physical and Philosophical Issues in the Bohr-Einstein Debate', pp. 8 f. of preprint.

5

Bohr's Philosophy of Physics

> A precondition to a full vindication of Bohr's philosophy is to fill the lacunae of his exposition and to transform his suggestions into a systematic and coherent world view.
>
> Abner Shimony

5.1. *On the Philosophy of Physics*

'I have been thinking a good deal [more] on the philosophy of physics' wrote Bohr to Hartree in 1929.[1] Indeed, Bohr gave a great deal of thought to the factors which made physics possible: the regularity of nature and the tantalizing elusiveness of nature's harmony, the character of our experience of nature, and the conditions for the unambiguous communication of our observations. In previous chapters we have examined Bohr's epistemology and its application to the problems raised by quantum theory. The task before us now incorporates some of the same issues, but viewed from a different perspective. I want to piece together Bohr's various remarks about some aspects of the nature of physics. The specific topics include discussions of objectivity and realism, the role of 'pictures' and symbols, the mutuality of analysis and synthesis, and the place of experiment and revision. Bohr made constant reference to these questions throughout his writings on the implications of atomic physics. This chapter, therefore, draws together and assesses what appear to be Bohr's chief preoccupations in a philosophy of physics.

A common if uncritical view of science portrays the physicist as someone who is able to observe the world dispassionately, who is detached from the world, and who is able to produce objective theories about the world. This view assumes that there is an external world which is the cause of the physicist's observations and which is responsible for the regularity in such observations. Further, it entails a correspondence theory of truth: that is, the formulae, concepts, and propositions employed in scientific discourse are taken to be true if they are in one-

[1] Letter to D. R. Hartree, 5 Jan. 1929, *BSC*: 11.

to-one correspondence with the systems of physical objects which constitute the external world. The scientific theories which the physicist produces are thus thought to represent and correspond to the hidden mechanisms underlying the observed behaviour of systems.

It will become clear that Bohr did not completely agree with this naïve view of science, although he did take such a view as implicit in our ordinary everyday descriptions of macroscopic external reality. The results of these investigations, however, do little to relieve the complexity of Bohr's thought. For the reader who has kept up so far, this will not come as a surprise. What characterizes his view of physics, once again, is the balancing, or, better still, the juggling of antithetical positions. In some passages, for example, Bohr relinquishes absolute distinctions between subject and object or between observer and observed. Again, he does not come down firmly on any one side in debates about the relative priorities of word and world or of theory and observation. Yet on the other hand he also respects the significance of the distinctions.

Rather than blurring the classical disjunction between subject and object, he attempts to recast the usual account of the objectivity of science from a new perspective. In particular, while rejecting the strong objectivism of classical science, he wants to make his own claims for the objectivity of physics and the abiding reality of nature. His main interests, of course, are the limitation which the quantum of action imposes on the application of the usual concepts and conceptual frameworks, and, as a parallel case, the mutuality of subject and object in the formulation of descriptions of experience. Some might call this a transcendental holism. Bohr called it complementarity. In a radical rather than compromising manner, he holds the middle ground.

'The goal of science is to augment and order our experience,' Bohr often observed, 'and deepen our understanding of the nature of which we ourselves are part.'[2] The emphasis on 'our experience' should not be interpreted to mean that Bohr is abandoning realism and reducing physics to phenomenalism. He cannot argue for both phenomenalism and complementarity at the same time, except in the most subtle of ways. Nor does he wish to make the point that all experience of nature is subjective; rather, he is cautioning against the naïve view of science which classical physics, at least to some extent, supported. The classical

[2] See *APHK*, p. 88, and the speech to Lord Boyd Orr, 14 Dec. 1949, *MSS*: 19 respectively; see also *ATDN*, p. 1, and *Essays*, p. 10.

ideal of our taking a stance apart from the world as detached observers is, in Bohr's view, necessarily limited.

In classical physics, which only operates successfully on a scale in which the effect of the quantum of action can be neglected, it is legitimate to assume a sharp distinction between the objects under discussion and the observer. Classical theory, indeed, provides an account of past and future states of physical systems independent of our observation of them. Bohr observes:

> Above all, the principles of Newtonian mechanics meant a far-reaching clarification of the problem of cause and effect by permitting, from the state of a physical system defined at a given instant by measurable quantities, the prediction of its state at any subsequent time. It is well known how a deterministic or causal account of this kind led to the mechanical conception of nature and came to stand as an ideal of scientific explanation in all domains of knowledge, irrespective of the way knowledge is obtained. . . .
>
> Within its large field of application, classical mechanics presents an objective description in the sense that it is based on a well-defined use of pictures and ideas referring to the events of daily life.[3]

In Bohr's view, the detached objectivism of the causal and mechanical account of nature was, at best, valid only on the macroscopic scale. Despite the certitude which mathematical treatments lent to physics, conceptual accounts were still dependent for their meaning on the circumstances of the subjects who used them. Once again, also, we find Bohr hinting at the close connection between the conditions for the success of ordinary language and for the objectivity of classical physics: both have to do with the well-defined use of re-identifiable particulars.

In his review of the ethos of classical science, Bohr moves on to query the 'idealization' of Newtonian physics beyond the range of experience to which our elementary concepts are adapted. He argued that the very recognition of the significance of the standpoint of the observer eventually produced the reshaping of Newtonian physics at the hands of Einstein.[4] 'The feature which characterizes the so-called exact sciences is', says Bohr, 'the attempt to attain to uniqueness by avoiding all reference to the perceiving subject.'[5] If Einstein's studies of relativity implicitly questioned this feature, then the discovery of quantum mechanics seemed, to Bohr's mind, to show up the limitations of classical physics once and for all.

[3] *APHK*, p. 69.
[4] Ibid., pp. 69 f.
[5] *ATDN*, pp. 96 f.

At the quantum level we are unable to claim sharp distinctions between observing system and quantum system, nor can we apply pictorial representations with certainty beyond the appropriate context for such descriptive terms. The discovery of the quantum thus 'made it evident that the classical physical theories are idealizations valid only in the description of phenomena in the analysis of which all actions are sufficiently large to permit the neglect of the quantum. While this condition is amply fulfilled in phenomena on the ordinary scale, we meet in atomic phenomena regularities of quite a new kind, defying deterministic pictorial description.'[6] Bohr's philosophy of physics, therefore, attempts to defend a new account of objectivity and realism in the broader context of quantum theory and the limitation to the univocal application of descriptive concepts.

What follows in this chapter, therefore, are some initial moves 'to fill the lacunae' of Bohr's exposition and to 'transform his suggestions into a systematic and coherent world view'.[7]

5.2. *Objectivity and Realism*

In classical physics, according to Bohr, there was a strong connection between the provision of a determinist causal-mechanical account of nature and the claims for absolute objectivity. In classical physics the objectivity of science was geared to the success of our predictions and our ability to describe the state of the physical system at any time, whether or not we observed it. The system was, if you like, patently subject-independent. Given this pattern, the rationale behind classical physics can be seen to be related to the criteria for successful linguistic identification of experienced particulars in external reality: our ability to re-identify objects which are external to us provides a basis for primitive descriptive concepts and implicitly introduces a space-time framework for our experience.[8] Even in classical physics, however, the concepts employed to describe any system are created by the subject. One does not discover concepts in nature. Bohr's account of objectivity thus came to rest on the conditions which apply to the proper use of concepts.

[6] *APHK*, p. 71.

[7] See A. Shimony, 'Physical and Philosophical Issues in the Bohr-Einstein Debate', Proceedings of the Bohr Centenary Symposium of the American Academy of Arts and Sciences (forthcoming), p. 22 of preprint.

[8] More will be said about this point in Chapter 7, where reference will be made to Strawson's arguments in his *Individuals*.

In Bohr's philosophy of physics objectivity was not to be attributed to the unlocking of determinist or causal secrets and the consequent establishment of a totally self-regulating and independent system in external reality. In the final analysis, objectivity lay in the refinement of unambiguous communication based on a well-defined use of pictures and ideas referring to the events of daily life.

This claim can be recognized as the basis of Bohr's defence against Einstein's various challenges. But Bohr's criterion of objectivity is not merely a linguistic one, nor can it be squarely equated with a positivist stance. His stringent demands for unambiguous appeal to classical 'pictures' are a plea to respect the limitations, as well as the strengths, of any apparatus employed in physics. Quantum physics demands the use of apparatus which can only measure classical properties like wavelength and momentum. To say that the objectivity of quantum physics is restricted to these individual observations is not, therefore, to say that the observation creates the physical reality, but rather to acknowledge the bounds of classical observation.

What is Bohr's criterion for the objectivity of physics? Physics is objective if it provides unambiguous information. Another way of putting this is to say that there is nothing that can be deemed 'subjective' about marks on photographic plates and readings of spectrometers: all unambiguous information 'concerning atomic objects is derived from the permanent marks . . . left on the bodies which define the experimental conditions. . . . The description of atomic phenomena has in these respects a perfectly objective character, in the sense that no explicit reference is made to any individual observer . . .'[9] Bohr's notion of objectivity differs from the classical account, however, in that he stresses that our descriptions of nature are not descriptions of independently existing realities, but descriptions of *our encounters with* such realities. Thus his position includes a recognition of the holism of subject and object, since subject and object can only be separated arbitrarily. Hence, in the light of quantum physics, 'this "objectivity" of physical observations becomes particularly suited to emphasize the subjective character of all experience'.[10]

Nature was very real for Bohr. It was not a dream.[11] And if you wished to say that our experience of nature was a dream, it would still

[9] *Essays*, p. 3; see also *APHK*, p. 68, and section 2.9. above.

[10] *ATDN*, p. 1; see also *Essays*, pp. 12 f.

[11] See Weizsäcker's account of Bohr's rejection of positivism and dreaming in *The Unity of Nature* (New York, Farrar Strauss Giroux, 1980), 183 f., and Bohr's disavowal

be a dream in which we had learnt something and could learn more.[12] Bohr's focus is obviously not so much on what 'is' as on what can be known objectively; and what can be known objectively, we can assume, had 'real' status for Bohr. But nature could not be known objectively at the quantum level if by 'objectivity' one meant knowledge of the past and future behaviour of quantum systems and direct pictorial descriptions of them.

The 'general epistemological lesson' which Bohr found to be instantiated in quantum theory was that the idealized objectivity of classical physics was itself open to critique: any descriptive language belongs to the framework of the subject, and object and subject cannot be so sharply distinguished. Oskar Klein recalls Bohr's point of view: 'In this connection he [Bohr] chose as a particularly simple example the use of a stick when trying to find one's way in a dark room. Here the dividing line between subject and object is placed at its end, when the stick is grasped firmly, while, when it is loosely held, the stick appears as an object.'[13] Thus Bohr moves to establish a different account of objectivity. Quantum physicists, restricted to describing experiments in terms of the whole apparatus-system interaction, find themselves in the same position as people trying to describe in detail the totality of their mental activity, 'since the perceiving subject also belongs to our mental content'.[14]

So far as I can tell, Bohr does not offer a precise account of knowledge and understanding. He does not attempt to describe the structure of knowing or to establish a foundational epistemology. In so far as he is concerned here with the circumstances of language usage, Bohr's approach has something in common with pragmatics.[15] But he also

of positivism in his essay, 'On the Notions of Causality and Complementarity', in *Dialectica* 2 (1948), 312-19.

[12] Weizsäcker, *The Unity of Nature*, p. 184, quotes a remark made by Bohr to E. Teller when the latter argued that Bohr was wrong to claim that 'classical concepts' would forever dominate our way of expressing sense experience: 'You might as well say that we are not sitting here, drinking tea,' said Bohr, 'but that we are just dreaming all this.' And Kalckar remembers that Bohr once said: 'if it had all been a dream, we would have learnt something nevertheless!'

[13] O. Klein, 'Glimpses of Niels Bohr as Scientist and Thinker', in S. Rozental (ed.), *Niels Bohr: His Life and Work* (Amsterdam, North-Holland, 1967), 93; see also *ATDN*, p. 99.

[14] *ATDN*, p. 96.

[15] Shimony makes much the same point in 'Physical and Philosophical Issues in the Bohr-Einstein Debate' with reference to A. Peterson's interpretation of Bohr's philosophy in *Quantum Physics and the Philosophical Tradition* (Cambridge Mass., MIT Press, 1968), 63, 157, 188.

outlines symptoms and consequences of knowledge-claims: scientific knowledge rewards us with 'a wider view and a greater power to correlate', thus satisfying 'our desire for an all-embracing way of looking at life in its multifarious aspects'.[16] Yet, in a key essay on the 'Unity of Knowledge' Bohr side-steps any definition of the term 'knowledge' and reduces the question of knowledge to the problem of objective description of experience: 'The main point to realize is that all knowledge presents itself within a conceptual framework adapted to account for previous experience. . . . When speaking of a conceptual framework, we refer merely to the unambiguous logical representation of relations between experiences.'[17]

'The relation between subject and object', insists Bohr, 'forms the core of the problem of knowledge.'[18] Again knowledge, including knowledge obtained in physics, is betokened by unambiguous communication, and given that attention is paid to the conditions which permit unambiguous communication, such knowledge cannot be described as purely subjective. Unambiguous communication entails, and is entailed by, objective description. A condition which applies to any knowledge-claim in quantum physics, therefore, is the recognition of the wholeness of observing system and observed system and the elimination of ambiguity in observation-reports. Quantum physics has forced an acknowledgement of the impossibility of claiming the strong objectivity traditionally espoused in classical physics. In this regard Bohr stands among contemporary philosophers of science, of which more will be said below.

Another way of approaching Bohr's occupation of the middle ground lies in his remarks on the undesirable extremes of materialism and mysticism in physics. The strong objectivism of classical physics, Bohr seems to be saying, is linked with a mechanist-determinist view of nature. The disowning of any objectivity, on the other hand, leaves us with little more than firmly held personal beliefs: if we abandon experimental evidence we end up in a mystical world of theory. But Bohr will have nothing to do with either of these extremes. He rejects 'a mysticism which is contrary to the spirit of natural science' as readily as he does the supposedly strong objectivity of classical science.[19] 'Our only way of avoiding the extremes of materialism and mysticism', he once wrote, 'is the never-ending endeavour to balance analysis and

[16] *APHK*, pp. 5 f., 80.
[17] Ibid., pp. 67 f.
[18] *ATDN*, p. 117. [19] Ibid., p. 15.

synthesis.'[20] Neither theory nor fact has privileged status. Instead, science is engaged in a to-and-fro process which rests on the assumptions that the world exists and that we can communicate unambiguously about that world.

So, objectivity is not considered in terms of a correspondence between the way the world is and the way we think the world is. For Bohr, it seems that objectivity in physics is not only related to coherence and consistency within theories but also to the successful anchorage of theories in observations: 'there could be no other way to deem a logically consistent mathematical formalism as inadequate than by demonstrating the departure of its consequences from experience or by proving that its predictions did not exhaust the possibilities of observation'.[21] If objectivity has little to do with an accurate 'inner-outer' match between concept and object, and having argued for the limitations which impose conditions upon our use of concepts, then Bohr can only refer to a coherence and consistency between theory and observation as a condition for claiming objectivity in knowledge.[22] Such judgements are also open to revision as new evidence comes to hand.

Bohr thus implicitly makes a distinction between two kinds of descriptions of experience. Subjective descriptions are those in which no clear external points of reference are available; objective descriptions, on the other hand, are those in which external points of reference, such as re-identifiable macroscopic particulars and measuring apparatus, are available. Quantum physics is truly objective because such points of reference are available. This does not mean that Bohr was a 'macro-realist', who regarded large everyday objects as real and everything else as real only in so far as it could be related to such objects. The reference to the macroscopic is a reference to the conditions for unambiguous communication, since such conditions are crucial to Bohr's understanding of objectivity.

The notion of the objectivity of classical physics may be understood to rest on similar suppositions. For many, however, classical objectivity also includes the notion of an absolute separation between object and subject so that observation is deemed to be of an independent reality. Bohr could not accept this, as we have seen, and he was therefore forced to provide a new account of realism in physics.

[20] N. Bohr, 'Analysis and Synthesis in Science', in the *International Encyclopedia of Unified Science I* 1 (Chicago, University of Chicago Press, 1938), 28.

[21] *APHK*, p. 57.

[22] For Bohr's appeals to 'consistency' see *TSAC*, p. 126, *NBCW* 4, p. 328, *ATDN*, pp. 7, 10, 55, 77, *APHK*, pp. 19, 37, 59, 73, 74, 80, and *Essays*, p. 6.

'Realist' philosophies of science embrace the view that scientific theories describe observed phenomena as the consequence of objects which exist independently of the observer. Einstein, who might be termed a 'classical realist', tended to believe that the terms employed in the theory describing the state of a physical system also characterized the properties of the independent objects whose observed behaviour suggested and confirmed the theory. While Bohr obviously wished to defend the objectivity of science, his position once again appears 'phenomenalist' rather than 'realist', for, according to the phenomenalists, physical theory merely gives an acount of what the subject observes rather than of an underlying objective reality.

But Bohr did not mean to make a Kantian distinction between things-in-themselves and phenomena when he stressed the role of the observer in quantum physics. His usage of the word 'phenomenon', as noted in Chapter 2, was designed to refer to the whole interaction between object and instrument.[23] In quantum physics the only reality with which we can deal unambiguously and consistently is the object-apparatus whole. This reminds us once again of Bohr's objections to the EPR argument and his stress on our inability to make sharp distinctions between 'the behaviour of the objects themselves and their interaction with the measuring instruments'.[24] As Weizsäcker put it, for Bohr the objects are not *behind* the phenomena, but *in* the phenomena.[25]

Bohr equates objective description with a 'well-defined use of pictures and ideas referring to the events of daily life'.[26] The key word here, it seems to me, is 'referring'. Rather than take up the elements of positivism or of inner-outer correspondence in this remark, the significance of 'referring' ought to be examined. *Our* participation in *events* which are re-identifiable is the basis for Bohr's affirmation of the reality of our world and his rejection of dreaming. It is a fact that we can repeatedly and unambiguously refer to particular aspects of our world. This, in brief, would have been Bohr's response to the sceptical arguments of Hume or Berkeley. He did not advocate either phenomenalism or idealism, because the very possibility of a subject being able to communicate about ordinary experiences with consistency implied the regularity of an independently existing real world.

[23] See 'The Causality Problem in Atomic Physics', p. 24, and section 2.9. above, for a discussion of the problem of observation in quantum physics.

[24] *APHK*, p. 61.

[25] See Weizsäcker, *The Unity of Nature*, p. 185.

[26] *APHK*, p. 69.

If Bohr was a realist, what kind of realism did he espouse? Shimony has argued that Bohr was neither a naïve realist, nor a macro-realist, nor a critical realist.[27] The first two claims are relatively incontestable, but something should be said about Bohr and critical realism. Shimony links his understanding of critical realism to the Kantian critique: knowledge of things cannot be obtained directly and naïvely through perception alone, but rather indirectly and inferentially. And, Shimony argues, Bohr's constant appeal to the necessity of using complementary experimental arrangements and the impossibility of using a single picture 'can be construed as a reason for rejecting in principle the feasibility of the critical realism'. In so far as this was not Bohr's approach to realism, it is correct to say that he was not a critical realist in the post-Kantian sense. But if by critical realism one means that we can discover the real through paying attention to the conditions for communication about experience, then Bohr could be described as a critical realist. It remains true, none the less, that Bohr did not pay as much attention to questions about the reality of 'things' as he did to the issue of our knowledge of things.

Folse describes Bohr as adopting 'causal entity realism': 'Bohr accepts realism because atomic theory can be used to design experiments which produce phenomena that are explained as the causal effect of the behaviour of atomic entities'.[28] This is also partly true, but it is a rather indirect way of making only one of many points. Indeed, any effort to describe Bohr's realism is open to this criticism: Bohr avoids a direct account of realism precisely because he is broadening the notion of the objectivity of physics to such a degree that reference to realism may confuse his path with vestiges of the very ideas he is trying to avoid. Instead of categorizing realism, therefore, I would like for the time being to describe Bohr's stance as 'relative realism'.[29] This is only to make the point that Bohr is a realist, perhaps with elements of critical and causal entity realism in his outlook, but that his focus is on the

[27] Shimony, 'Physical and Philosophical Issues in The Bohr-Einstein Debate', pp. 16 ff. of preprint.

[28] H. Folse, 'Niels Bohr: Complementarity and Realism', p. 4 of preprint (to be published by the Philosophy of Science Association). Folse here refers to I. Hacking, *Representing and Intervening* (Cambridge, Cambridge University Press, 1983), 274. See also H. Folse, *The Philosophy of Niels Bohr: The Framework of Complementarity* (Amsterdam, North-Holland, 1985), 222 ff.

[29] This term is not chosen with Einstein's relativity theory in mind, but rather with reference to Strawson's recent observations in *Skepticism and Naturalism* (London, Methuen, 1985), 44, defending 'a certain ultimate relativity in our conception of the real'.

conditions of unambiguous communication in physics rather than on how the theory represents the reality.

Yet if we accept that Bohr was some kind of realist, then we must consider the question of the intrinsic properties of the atomic objects. Realism in quantum theory leaves us with the very difficult problem of explaining what physicists call the reduction of the wave packet. A stream of electrons diffracted through a narrow aperture, as in Figure 4.1 above, behaves as a wave until the moment it strikes a screen, when the spreading wave packet suddenly registers as a particle. The indefatigable questioner will always want to know what the reality of the electron stream is. Einstein certainly thought that physics was about knowing reality independent of its observation. By stating that the formalism of quantum theory is concerned only with what happens when observations are made, then Bohr is restricting Einstein's vision of physics. To some he may even appear as an anti-realist.

Furthermore, the idea that the quantum formalism describes an interaction between macroscopic and quantum objects leads to rather paradoxical situations, best illustrated by the case known as Schrödinger's cat. Schrödinger proposed the following thought-experiment: a cat is placed in a kind of electric chair in a steel box which also contains a small radioactive source and a Geiger counter. There is an even chance that the radioactive source will or will not disintegrate, and if it does, the Geiger counter will register its emission and trigger a signal that electrocutes the cat. Imagine that we do not open the box for an hour; we do not know whether the cat is alive or dead. Indeed, the quantum formalism for this system suggests that, according to Bohr, opening the box and looking at the cat will throw the system into 'cat alive' or 'cat dead'.

This does indeed seem ridiculous: the cat is obviously going to be alive or dead whether or not we look at it. Bohr's opponents would suggest, therefore, that there are in fact real properties to be observed 'behind' the system-observation interaction. Bohr can reply, however, that in such a case the cat itself is the macroscopic observer and that the reduction of the wave packet occurs when the cat feels the electric jolt, if it takes place.

Because Bohr rejected the kind of realism that rested on a correspondence between the theoretical formalism and physical objects, he found the case of Schrödinger's cat rather trivial. What the case illustrates, however, is that the notion of sharply defined macroscopic systems has to be introduced from outside quantum theory itself. I do not think

Bohr would be too troubled by this, for he never showed great interest in self-contained axiomatic foundations for the theory. On the other hand, it does demonstrate how essential the framework of complementarity is for an understanding of the paradoxical connections between quantum physics and classical concepts.

A causal mechanical framework can be used successfully at the level of ordinary and everyday descriptions: descriptive concepts we apply to the independent objects which are the cause of our observations are univocal. Einstein wished to maintain such a framework, broadened by a field-theoretical ontology, even in the realm of quantum physics, but, as we have seen, he encountered the difficulties of indeterminacy. Bohr's advocacy of complementarity was not an abandonment of realism but an attempt to provide a more general framework for the description of nature; maintaining objectivity and realism would be at the cost of reappraising the status of our usual physical concepts. In particular, we would need to learn that such concepts were symbols, pictures, and idealizations when applied to quantum events.

5.3. *Symbols and Pictures*

If physics is, for Bohr, about the ordering of our experience, then it also entails the use of those symbols and pictures which we deem to represent the objects of our experience. This is true not only of ordinary picture-words like 'wave' and 'particle', but also of particular significance in the theoretical and mathematical models which have contributed so much to the advance of science. In as much as physicists engage in discourse about the physical world, questions about the nature of language are important in any philosophy of physics. In the classical view, moreover, in which the concepts and formulae of physical theory are taken to correspond to and represent real independent objects, there is an underlying assumption of a universal scientific language which is subject-independent.

Bohr's approach challenges such assumptions. In this section I intend to delve into Bohr's various reflections on the use of language in physics and the problem of representing reality in terms of symbols, concepts, pictures, abstractions, and frameworks. Though Bohr is sometimes inconsistent in his use of these key terms, the underlying point remains the same: there is no universal and independent scientific language; rather, we can only use descriptive concepts objectively when we include an account of the context in which they are applied.

The links which Bohr discerns between ordinary language and classical mechanics will also become more apparent here. Just as a descriptive concept like the word 'tree' draws its significance from our ability to identify such objects in the realm of ordinary experience, so also, for Bohr, the descriptive terms of classical physics have similar provenance. Thus, even a sophisticated term like 'momentum' is anchored in our observation of exchange of energy between moving bodies. How such 'picture'-words are to be applied to quantum events, therefore, raises questions about macroscopic apparatus and the scope of our powers for forming concepts.

In Bohr's earlier work the term 'symbol' has great currency . In his later writings, however, he prefers to use terms like 'pictorial concepts' to focus his discussion of the limitations to language which condition our descriptions of physical reality. We will start with a discussion of his usage of 'symbol', therefore, and then move on to a consideration of the implications of 'pictures' in language.

Bohr's appeal to 'symbol' is frequent, but its meaning shifts from very technical applications in mathematical formulae to the more general referring role of ordinary descriptive words. With reference to quantum theory, he states 'that the symbolical garb of the methods in question closely corresponds to the fundamentally unvisualizable character of the problems concerned'.[30] What is 'unvisualizable', in other words, can only be denoted symbolically. Again, he qualifies the difference between 'symbols' and 'visualizables': classical terms which also convey a 'picture', like 'wave' and 'particle', are described as merely symbolic when applied to quantum events.[31] When 'wave' and 'particle' are applied to quantum events, however, Bohr normally refers to such applications as 'abstractions' or 'idealizations'. This is not meant to reflect on the questionable reality of the quantum world as to respect the limits to our unambiguous communication and the appropriate ways of claiming objectivity.

In Bohr's writings in the late 1920s 'symbol' is a keyword. As such, it is associated with a renunciation of visualization in exactly the same way that quantum theory entails a renunciation of classical physics. In his crucial addresses of 1927, for example, Bohr uses 'symbol' and its derivatives at least ten times. His remarks, nevertheless, fall into several different categories.

To begin with, Bohr cites several cases in which quantum theory provides us with symbols which specify and characterize our situation

[30] *ATDN*, p. 12. [31] Ibid., p. 17.

in the world. For example, Planck's quantum of action 'symbolized' 'an essential discontinuity, or rather individuality, completely foreign to the classical theories'; again, the co-ordination of spatio-temporal and causal frameworks in classical physics was seen as 'symbolizing the idealization of observation and definition respectively'; and finally, the new indeterminacy relations 'may be regarded as a simple symbolical expression for the complementary nature of the space-time description and the claims of causality'.[32] Here 'symbol' is almost equivalent to 'emblem', but it also expresses the fact that a total change of gear must occur in our thinking as we shift from classical to quantum physics.

In a second set of applications, Bohr uses 'symbolic' to describe the relationship between classical concepts and quantum events. Thus 'we symbolize' the interactions between quantum objects 'by the abstractions of isolated particles and radiation'; again, he speaks of 'the "individuals" symbolized through the conception of light-quanta'.[33] Bohr is alluding here to his argument that descriptive concepts from the everyday world can only be applied to the quantum realm at one remove, as it were, or as abstractions. If we pretend to talk about the reality of quantum objects independent of their observation, then it cannot be in the same manner that we talk about the properties of demonstrably independent macroscopic objects. Another way of putting this would be to say that the descriptive terms from classical physics can only be used analogically.

Finally, and most commonly, Bohr refers to the quantum formalism itself, the matrix-mechanics and wave-equations, as a 'symbolic method'. The symbolic character of these equations rests not just on the use of sophisticated mathematics, however, but also on the necessary failure of our ordinary classical concepts or visualizables:

> The symbolical character of Schrödinger's method appears not only from the circumstance that its simplicity . . . depends essentially upon the use of imaginary arithmetic quantities. But above all there can be no question of an immediate connection with our ordinary conceptions because the 'geometrical' problem represented by the wave equation is associated with the so-called co-ordinate space, the number of dimensions of which is . . . in general greater than the number of dimensions of ordinary space . . .
>
> On the whole, it would scarcely seem justifiable . . . to demand a visualization by means of ordinary space-time pictures.[34]

[32] 'The Quantum Postulate and the Recent Development of Atomic Theory', in *ATDN*, pp. 53, 55, 60.

[33] Ibid., pp. 69, 89.

[34] Ibid., pp. 76 f.; see also pp. 71, 73, 90.

Given that the wave-equation operates in multidimensional space, as well as for other reasons we have referred to, Bohr concluded that no direct correspondence could be claimed between the application of descriptive concepts in our familiar space-time frameworks and in the interpretation of the quantum formalism.

Whereas in classical physics a one-to-one correspondence is assumed to exist between the properties of objects in a physical system and the mathematical formula which links such properties, Bohr argues that the quantum formalism can only be regarded as representing such properties symbolically rather than univocally:

> The quantities which in classical physics are used to describe the state of a system are replaced in quantum-mechanical formalism by symbolic operators whose commutability is limited by rules containing the quantum. This implies that quantities such as positional coordinates and corresponding momentum components of particles cannot simultaneously be ascribed definite values. . . . In addition, this generalization permitted a consequent formulation of the regularities which limit the individuality of identical particles and which, like the quantum itself, cannot be expressed in terms of usual physical pictures.[35]

In the light of this third set of usages of the term 'symbol', focus can now be shifted to the connection between the breakdown of the applicability of our 'usual physical pictures' and the role of symbols in quantum theory.

Bohr's writings of 1928, particularly his correspondence, reveal his continuing struggle to clarify the meanings of the terms 'symbolical' and 'visualizable'. In March of that year he wrote to Dirac: 'I think, we can not too strongly emphasize the inadequacies of our ordinary perception when dealing with quantum problems.'[36] A couple of months later, this time to Darwin, Bohr displayed the same preoccupation: 'I do not think it is possible to stress the symbolic character of the quantum theoretical methods too much.'[37] Exactly what Bohr had in mind is best exemplified in his letter to Christian Møller, a young physicist who had asked him to clarify his notions of 'symbol' and 'visualizable'. Bohr wrote in reply:

> I have only thought to stress the fact that our circumstantial use of the symbols of classical theory in quantum theory does not permit us to disregard the great difference there is between those theories; in particular, it calls for the greatest

[35] *APHK*, p. 87.
[36] Letter to Dirac, 24 Mar. 1928, *BSC* : 9, *NBCW* 6, p. 45.
[37] Letter to Darwin, 9 May 1928, *BSC* : 9.

carefulness in the use of those forms of perception to which the classical symbols are connected. . . . When one thinks of the wave theory it is, however, just its 'visualizability' which is both its strength and its weakness; in stressing the symbolic character of the treatment I have sought to remind one of the great difference (brought about by the quantum postulate) from the classical theories, a difference which has not always been kept in mind.[38]

Two comments can be made here. First, our usage of words and images ought to be recognized as 'circumstantial': words do not have any privileged status in themselves and, therefore, their proper usage entails some sort of reference to the appropriate circumstances. This, put in another way, is Bohr's appeal to the conditions for unambiguous communication. Secondly, whereas classical physics entails a use of words and symbols in which some elements of discourse are supposed to have direct reference to objects or interactions in the world, quantum physics is greatly different. Because of the discontinuity introduced by the quantum, the visualizability which is the strength of classical theory is denied to quantum physics. Whatever properties are observed in the observations of quantum events, Bohr persistently argued, cannot be applied directly to the quantum object itself independent of the context of observation. A rose remains a rose, Bohr would say, no matter how we look at it and whether we look at it or not. To speak of an electron as a particle outside the context of a scattering experiment, however, is to employ the term 'particle' as an abstraction.

From this point of view, both 'symbol' and 'visualizable' are descriptions of our makeshift bridges across the gap between the everyday view of reality and the quantum view of reality. Because of the change in circumstances in our use of these terms, whether descriptive or mathematical, our entire discourse must be regarded as symbolic. Once again Bohr is bringing us back to face the problems of the relationship between theory and observation, concept and reality.

In his later writings Bohr is quite explicit about the overlap between mathematical and linguistic symbols: 'For our theme it is important to realize that the definition of mathematical symbols and operations is based on simple logical use of common language.' Thus mathematics is seen by Bohr as a refinement of general language where ordinary verbal expression is too cumbersome. Linking mathematical procedures to experiments depending on classical observation and description, Bohr wrote in the same essay from 1958: 'In the treatment of atomic

[38] Letter to Møller, 14 June 1928, *BSC*: 14, from the Danish, following Murdoch's translation.

problems, actual calculations are most conveniently carried out with the help of a Schrödinger state function. . . . It must be recognized, however, that we are here dealing with a purely symbolic procedure, the unambiguous physical interpretation of which in the last resort requires a reference to a complete experimental arrangement.'[39] Here Bohr adds nuances to the meaning of symbol as a necessarily conditioned representation of an elusive reality. The only quantum reality which can be accounted for unambiguously is that which is observed in the interaction between measuring system and quantum object. Moreover, further specification of one aspect of that reality will entail the neglect of other aspects.

There remains yet another feature of Bohr's usage of 'symbol'. Not only is a quantum account 'symbolical', in so far as it depends on classical terms and apparatus, but a classical description is also symbolic, in as much as the everyday picture of reality is a limited and incomplete ordering of experience. Heisenberg has remarked on Bohr's predilection for parables,[40] and Kramers has recalled Bohr's reference to 'caricature': 'Bohr has expressed himself in discussions somewhat as follows: classical physics and the quantum theory, taken as descriptions of nature, are both caricatures; they allow us, so to speak, to asymptotically represent actual events in two extreme regions of phenomena.'[41] In other words, although descriptive terms can be used univocally in the classical framework, it should be remembered that the causal and space-time frameworks are of only limited validity; accurate representation of reality in classical physics depends on the assumption of a causal and space-time framework, and this in itself is an abstraction. Quantum theory in fact provides a more general account of nature. The paradox is that each theory needs the other: on the one hand quantum theory requires the descriptive concepts of everyday experience and classical physics; and on the other hand quantum mechanics succeeds where classical theory fails. The two approaches, one continuous and one discontinous, are compatible only in their convergence.

Although Bohr does use the word 'symbol' in specific ways, such as in a description of the contents of the Schrödinger wave function, these do constitute special cases. His more general concern is to emphasize the impossibility of the claim that 'words' can represent 'reality'

[39] *Essays*, p. 5; see also p. 9.

[40] See W. Heisenberg, *Physics and Beyond* (London, Allen and Unwin, 1971), 210.

[41] H. Kramers, as quoted by Feyerabend in his *Realism, Rationalism and Scientific Method* (Cambridge, Cambridge University Press, 1981), 253 n. 15.

precisely and completely. Instead, as far as Bohr is concerned, words orientate us in reality by their immediate or derived reference to demonstrable circumstance. Given that the meaning of words is circumstantial, he allows for a continuity between 'word' and 'object' as much as he does for the separation between word and world which a mirroring theory of knowledge would entail. Whereas the classical physics of Newton and the unified physics of Einstein aspired to a precise replication of reality through well-defined mathematical theory, Bohr's account runs quite the other way.

When 'word' is seen as continuous with reality rather than standing apart from it, then one is doing something equivalent to the abolition of 'inner-outer' distinctions. Bohr's use of the terms 'symbol' and 'picture' is not only about the representation of reality, but also about the limits and conditions which apply to our representation of reality. Instead of appealing to a sharp distinction between 'word' and 'world', Bohr operates more as a holist. This is not to abolish all distinctions between picture and reality, but to recognize that there is a fluidity between our words and our world: each discloses the other just as much as it warrants the other.

In Bohr's later discussions of these issues he refers more to 'pictures' than to 'symbols'. In his essay from 1955, 'Atoms and Human Knowledge', for example, 'symbolic' occurs only once, whereas 'picture' and its derivatives occurs some nine times. His quest to establish a new kind of objectivity in quantum physics, as he puts it in this essay, brought him to 'a thorough analysis of the scope of pictorial concepts'.[42] The 'decisive point', as we have seen, is that 'the description of the experimental arrangement and the recording of observations must be given in plain language, suitably refined by the usual physical terminology. This is a simple logical demand, since by the word "experiment" we can only mean a procedure regarding which we are able to communicate to others what we have done and what we have learnt.'[43] The experimental basis for the success of classical physics, that is, is the same as the requirement of some form of demonstrability in the acquisition of our descriptive concepts.

Bohr's discussion of measurement and observation in quantum physics constantly returned to the fact that the quantum discontinuity and the indeterminacy relations 'not only set a limit to the *extent* of the

[42] *APHK*, p. 86.
[43] *Essays*, p. 3; see also the discussion of measurement in section 2.9. above.

information obtainable by measurements, but they also set a limit to the *meaning* which we may attribute to such information'.[44] In other words, the 'meaning' which our descriptive concepts carry is limited to demonstrable effects on apparatus. This was why Bohr took such pains to illustrate his controversial thought-experiments with screens, clocks, spring balances, rigid frameworks, and the like. Here it was that pictorial terms had their cash value.

In one of his last essays, entitled 'The Unity of Human Knowledge', Bohr brings together his arguments about the objectivity of physics and the limiting role of language:

> Indeed, from our present standpoint, physics is to be regarded not so much as the study of something *a priori* given, but rather as the development of methods for ordering and surveying human experience. In this respect our task must be to account for such experience in a manner independent of individual subjective judgement and therefore objective in the sense that it can be unambiguously communicated in common human language.[45]

Once again Bohr's mode of thinking rests on simply grasping the point of, and elucidating what is constitutive of, a particular activity. If science is about the unambiguous communication of experience, then it is concerned with the use of words which have unambiguous meaning and which, therefore, must be defined in terms of our shared awareness of relatively fixed circumstances in the external world.

At the same time, especially in his later writings, Bohr makes it clear that he is not intending to reduce physics to a thoroughly relativist, if self-consistent, language game. Over and over again, even as he qualifies our ability to describe them, he refers to the 'atomic objects' at the heart of physical reality.[46] The analysis of the scope of our pictorial concepts entitles us to regard their univocal application as limited and to employ complementary descriptions. Complementarity thus becomes the broader framework for the possibility of objectivity in quantum physics which Bohr seeks to defend.

So far little has been said about the *methods* for ordering and surveying human experience. We shall now turn our attention first to Bohr's view of 'the balancing of analysis and synthesis', and then to the role of experiment, theory, and revision.

[44] *ATDN*, p. 18, Bohr's emphasis.
[45] *Essays*, p. 10.
[46] See, for example, 'Quantum Physics and Philosophy', in *Essays*, pp. 3-6.

5.4. *Balancing Analysis and Synthesis*

Bohr's choice of terminology is both deliberate and idiosyncratic. Terms like 'phenomenon' and 'individuality', for example, are given special meanings and recur throughout his writings. Equally intriguing and surely instructive are the words which Bohr habitually employs to describe the methods for increasing human knowledge, namely 'analysis and synthesis'. Establishing what Bohr means by these terms and seeing how he uses them illuminates an unusual aspect of his understanding of physics: Bohr warns against physics being preoccupied only with the discovery of mechanisms in nature.

We have seen above how Bohr defined the goal of science in terms of the augmenting and ordering of our experience: 'The very essence of scientific explanation', says Bohr, is 'the analysis of more complex phenomena into simpler ones.'[47] 'Analysis' refers to splitting both experimental data and theoretical concepts into components. Thus 'concepts', 'word', 'psychic experiences', and 'pictorial concepts' all provide 'content' for analysis.[48] We can also engage in 'closer analysis in experimental terms'.[49] This was all part of the 'augmenting' of experience. Finally, and chiefly exemplified in Heisenberg's work, one could also engage in mathematical analysis.[50] Analysis thus signifies movement from the complex whole to the single elements. Bohr even sometimes used 'description' as a synonym for 'analysis' (and 'comprehension' as a synonym for 'synthesis').[51]

If such analysis could be experimental, mathematical, or even philosophical, Bohr was perhaps unique as a physicist in that he operated in all three areas at the same time. Heisenberg offers beautiful evidence of this:

> his [Bohr's] insight into the structure of the theory was not a result of a mathematical analysis of the basic assumptions, but rather of an intense occupation with the actual phenomena, such that it was possible for him to sense the relationship intuitively rather than derive them formally.
>
> Thus I understood: knowledge of nature was primarily obtained in this way, and only as the next step can one succeed in fixing one's knowledge in mathematical form and subjecting it to complete rational analysis. Bohr was

47 *APHK*, p. 3.

48 See respectively *Essays*, p. 3, *ATDN*, p. 20, 'Causality and Complementarity', p. 298, *APHK*, pp. 86, 98, and *Essays*, p. 12.

49 See *APHK*, pp. 3, 51.

50 See *Essays*, pp. 9, 78.

51 See *APHK*, pp. 74, 91.

primarily a philosopher, not a physicist, but he understood that natural philosophy in our day and age carries weight only if its every detail can be subjected to the inexorable test of experiment.[52]

Bohr's stress on coherence with experimental evidence here is another indicator of his realism. For Bohr the work of analysis is to get as close as possible, and as objectively as possible, to the detail of nature as we encounter it.

The 'ordering' of experience, on the other hand, is a more intuitive activity which has to do with models, theory, and formalism. It entails grasping the 'whole' rather than any one aspect of detail. And, such a 'whole', as Bohr worked, needed to take into account mathematical, epistemological, and observational aspects. Heisenberg has rightly identified Bohr's great gift for such intuition. Bohr called it *synthesis*. The framework of complementarity, as Bohr propounded it, is an example of such synthesis: 'In offering a frame wide enough to allow a harmonious synthesis of the peculiar regularities of atomic phenomena, the conception of complementarity may be regarded as a rational generalization of the very idea of causality.'[53] Elsewhere he describes complementarity as a 'rational synthesis'.[54] Just as he occasionally used 'description' as a synomym for 'analysis', so Bohr also used 'comprehension' and 'rational generalization' as synonyms for 'synthesis'. If analysis implied splitting apart, then synthesis had to do with drawing together. In Bohr's view, analysis and synthesis must go hand in hand in the progress of science.

So far there is nothing very surprising about Bohr's view of method: our usual way of learning about things is to take them apart and put them together again. What is enlightening, however, is Bohr's emphasis on the *balancing* of the two methods. As we shall see, Bohr thought too much emphasis on analysis was dangerous, for such an activity carried with it the covert notion that nature could ultimately be reduced to an independent mechanical system composed of small particle-like objects. Bohr was clearly opposed to such reductionism. Equally, he was cautious about overstressing theoretical synthesis, for that carried with it the temptation of a purely metaphysical and deductive view of nature, removed from the hard evidence of experience.

[52] Heisenberg, 'Quantum Theory and Its Interpretation', in Rozental, *Niels Bohr*, p. 95.

[53] N. Bohr, 'Atomic Physics and International Cooperation', *Proceedings of the American Philosophical Society* 91 (1947), 98.

[54] *APHK*, p. 19.

Bohr saw too much emphasis on analysis as leading to an exclusively materialist conception of nature. An unjustified extension of the notions of space and time and mechanical causation would also bring with it the possibility of a universal strong separation between known objects and the knowing subject. On the other hand, too much stress on synthesis was concomitant with a neglect of the real world and opened the way to subjectivism in the pejorative sense of the word. In 1937 Bohr wrote: 'in abandoning the causal description in atomic physics we are not concerned with a hasty assertion of the impossibility of comprehending the wealth of phenomena, but with a serious effort to account for the new types of laws here encountered in conformity with the general lesson of philosophy regarding the necessity of a balance between analysis and synthesis'.[55] A year later, in his article for the first volume of the *International Encyclopedia of Unified Science*, Bohr returned to this theme: 'even in science any arbitrary restriction implies the danger of prejudices, and . . . our only way of avoiding the extremes of materialism and mysticism is the never-ending endeavour to balance analysis and synthesis'.[56] The parallels which Bohr draws between 'materialism' and 'analysis', and between 'mysticism' and 'synthesis', are key indicators of his meaning here. For a man who has been identified by many with positivism, Bohr proves to be, as Holton puts it, the 'best example of a major spirit of science in our century refusing to favour one polar position or the other'.[57] It is not clear from these statements, however, just what arguments or reasons Bohr has in mind.

Bohr's talk about balancing analysis and synthesis, and therefore of balancing experiment and theory, reflects more than a passing phase in his thought. Fortunately his later writings provide some clues to his reasoning. In the second sentence of his Preface to *Atomic Physics and Human Knowledge* he writes: 'The theme of the papers is the epistemological lesson which the modern development of atomic physics has given us and its relevance for analysis and synthesis in many fields of human knowledge.'[58] His 'epistemological lesson', arising from reflection on the conditions for unambiguous communication, is now placed at the centre of his reasoning. Any experimental analysis, given the conditions for unambiguous description which objectivity requires, is limited by the point of view adopted by the observer.

[55] 'Causality and Complementarity', p. 294.

[56] 'Analysis and Synthesis in Science', p. 28.

[57] G. Holton, *The Scientific Imagination* (Cambridge, Cambridge University Press, 1978), 140.

[58] *APHK*, p. v; see also pp. 33, 63, 68, and *Essays*, p. 7.

Again, in his extensive review of his discussions with Einstein, Bohr places the same issue to the fore:

The question at issue has been whether the renunciation of a causal mode of description of atomic processes involved in the endeavours to cope with the situation should be regarded as a temporary departure from ideals to be ultimately revived *or whether* we are faced with an irrevocable step towards obtaining the proper harmony between analysis and synthesis of physical phenomena.[59]

In other words, the advent of quantum physics provided an opportunity to correct the misapprehension, commonly associated with classical physics, that an objective external reality could be grasped as such and replicated in well-defined theory which could be considered totally subject-independent.

The twin methods of analysis and synthesis have been scrutinized in recent philosophy of science in terms of the interplay between observation and theory. In order to report observation, for example, one has to employ theoretical terms; likewise, in order to posit theories one requires observational terms. Experience and theory are thus seen to be intertwined. This holist view of scientific method arose in opposition to the positivist and sense-datum theories of knowledge.

The concordance of analytic and synthetic methods has defied philosophers of science from Aristotle to Albertus Magnus, perhaps also from Reichenbach to Rorty. Total abandonment of the distinction between the two methods does not resolve the issue. One cannot simply dismiss the question without placing in jeopardy the subtlety of distinctive meanings. Each can be given its proper application, it seems, and yet the interdependence of the two quite contrary methods must also be allowed for. Bohr stresses the limitations of each method and thus also their individual incompleteness. Each method demands the other, he argues, since the quantum imposes limits on the range of experience and also reflects the impossibility of sharp distinction between object and subject: 'The viewpoint of "complementarity" does, indeed, in no way mean an arbitrary renunciation as regards the analysis of atomic phenomena, but is on the contrary the expression of a rational synthesis of the wealth of experience in this field, which exceeds the limits to which the application of the concept of causality is naturally confined.'[60] 'Complementarity' is yet again seen as denoting the non-

[59] *APHK*, p. 33, my emphasis.
[60] *APHK*, p. 19.

sequential or holist understanding of our experience and description of reality. The key to Bohr's world-view is neither experience nor analytic method, but more theoretical ordering or synthesis, which will offer a wider and more real perspective on that experience.

Bohr places limits on both analysis and synthesis. The individuality of quantum events cannot be analysed *ad infinitum* 'in the customary way of classical physics', because of the 'unavoidable interaction between atomic objects concerned and the measuring instruments'; and, likewise, the dream of a perfect synthesis in terms of unified conceptual structures can never become reality, since it 'rests on the assumption . . . that it is possible to discriminate between the behaviour of material objects and the question of their observation'.[61]

Bohr is convinced that we must stand in the middle ground, favouring neither method and yet employing both. While this might be frustrating and confusing for minds schooled in the possibility of strong objectivity and inexorable empirical method, Bohr finds that the situation in physics matches his general epistemological conclusions about our situation as actors and spectators in the world and the conditions which constitute our successful use of descriptive concepts.

5.5. *Experiment, Theory, and Revision*

Given the need for method to shift constantly between analysis and synthesis, how does science progress? Bohr appears to deny science any logical certitude arising out of specific methodology. And yet experimentation, understood in a particular way, is of primary importance. Bohr found a kindred spirit in Picasso: 'I do not look for things, I find them.'[62] Those who look for things, in other words, will only find what they are looking for; those who simply find, however, will find what is there to be found. An experiment should not be an imposition upon physical reality, seeking to squeeze it into a particular preconceived shape. Instead, the results of an experiment should be allowed to challenge preconceptions about reality. Experimentation alone, however, is not enough.

At the beginning of *Atomic Theory and the Description of Nature*, therefore, Bohr elaborates on the balance between experiment and theory, arriving at the necessity of constant openness and revision:

[61] Ibid.
[62] See M. Andersen's contribution to Rozental, *Niels Bohr*, p. 323.

> The task of science is both to extend the range of our experience and to reduce it to order, and this task presents various aspects, inseparably connected with each other. Only by experience itself do we come to recognize those laws which grant us a comprehensive view of the diversity of phenomena. As our knowledge becomes wider, we must always be prepared, therefore, to expect alterations in the points of view best suited for the ordering of our experience. In this connection we must remember . . . all new experience makes its appearance within the frame of our customary points of view . . .[63]

Bohr is making two points: theory must be open to revision by experiment; and, we must keep in mind that the experiment itself should not be confined by particular theoretical frameworks. Hence, implicitly, intuition and theory can open new doors for experimentation.

Bohr's definition of the term 'experiment', therefore, is, not surprisingly, elliptical. An experiment does yield information, but the information is conditioned by the conceptual frameworks within which the observer operates: 'by the word "experiment" we can only mean a procedure regarding which we are able to communicate to others what we have done and what we have learnt'.[64] In particular, experiments restrict observers to the use of classical concepts (meaning everyday language and its refinements), since 'the requirement of communicability requires that we can speak of well-defined experiences only within the framework of ordinary concepts'.[65] Quantum physics is the inevitable exemplar of this claim, and not an exceptional case. Experimental measurement, as discussed above, means nothing more than 'standardized comparison'.[66]

Such objectivity, as Bohr understands it, allows first of all for the verification of theory, even in quantum mechanics: 'the emphasis on permanent recordings under well-defined experimental conditions as the basis for a consistent interpretation of the quantal formalism corresponds to the presupposition, implicit in the classical physical account, that every step of the causal sequence of events in principle allows of verification.'[67]

Experience is recognized by Bohr as leading to the discovery of new laws as well as obliging the revision of old points of view. For Bohr,

63 *ATDN*, p. 1.

64 *Essays*, p. 3.

65 'Causality and Complementarity', p. 293.

66 *Essays*, p. 6. For a more detailed discussion of measurement and Bohr's general interpretation, see sections 2.9. and 3.9. above.

67 *Essays*, p. 6.

mathematical theory is wrong if it does not fit the facts, albeit that the facts themselves may have to be viewed critically, even if the theory itself is internally consistent:

In my opinion, there could be no other way to deem a logically consistent mathematical formalism as inadequate than by demonstrating the departure of its consequences from experience or by proving that its predictions did not exhaust the possibilities of observation.[68]

In the natural sciences proper . . . there can be no question of a strictly self-contained field of application of the logical principles, since we must continually count on the appearance of new facts, the inclusion of which within the compass of our earlier experience may require a revision of our fundamental concepts.[69]

Although Bohr himself performed very few experiments, his earlier writings repeatedly stress the importance of experimental evidence for the confirmation or refutation of the theory.[70]

The progress of physics in this century not only greatly expanded our powers of observation and our tools for experimentation: classical conceptions of physics, including determinism and mechanistic causality and continuity, were challenged by new theoretical frameworks. 'New experience', Bohr writes, 'has time after time demanded a reconsideration of our views.'[71] Yet despite giving this priority to the 'augmenting' of experience, Bohr is ultimately interested in its 'ordering', that is, in the theory. This is not to make a sharp distinction between experience and theory, but to take cognizance of mutual influences on scientific progress. For if experience provides the context for our questions and corrections, theory is also immediately entailed. Bohr's emphasis on experimental data is not a fatalist reductionism of theory to sense-perception. Rather, the experimental data already offers opportunities for finding more general connections and, in this sense, already entails a new theoretical framework. The experimental evidence is for the scientist what the block of marble is for the sculptor. Bohr does not look for things, he finds them!

Bohr's notion of theory involves more than the ordering of sense-data. This is partly related to his view of symbols and visualizables, discussed above, in which experience is seen to entail much more than either a

[68] *APHK*, p. 57.

[69] *ATDN*, p. 97. Note that this 'revision' is based on experience, and not on the a priori methods of revisionary metaphysics.

[70] See, for example, *TSAC*, pp. 4, 16, 60, *NBCW* 2, pp. 286, 298, and *NBCW* 3, p. 282.

[71] *APHK*, p. 32; see also p. 67.

raw feel or a mirror-imaging. Bohr implicitly rejects both these extremes. In quantum physics especially, the relationship between theory and observation becomes more acute. Here, according to Bohr's philosophy, the ordering of experience is raised to the more general level of symbolic abstraction. Whereas in classical physics the application of descriptive concepts was similar to the usage of descriptive concepts in ordinary language, in quantum physics there was no warrant for such a parallel. Hence that process of revision in the light of experiment, which Bohr took to characterize the progress of classical science, now evolved into a transformation of physics itself: the interplay between concrete experience and classical theory is elevated to a mutuality of 'phenomena' (meaning object-apparatus whole) and 'abstraction':

Still, it became more and more clear that, in order to obtain a consistent account of atomic phenomena, it was necessary to renounce even more the use of pictures and that a radical reformulation of the whole description was needed to provide room for all the features implied by the quantum of action. . . .

As in the formulation of relativity theory, adequate tools were found in highly developed mathematical abstractions.[72]

'Abstraction', however, should not be taken to mean a movement away from reality so much as an effort to indwell in reality in a new and more general manner by avoiding the limitations imposed by a classical 'picturing' of everyday circumstances.

Controlling the development of these more abstract syntheses raises further issues; since they now lie beyond the edge of direct observation, the fitting to the facts becomes more complicated. In this context, once again, Bohr returns to the importance of both inner consistency in the theory and coherence with experience:

In fact, all our knowledge concerning the internal properties of atoms is derived from experiments on their radiation or collision reactions, such that the interpretation of experimental facts ultimately depends on the abstractions of radiation in free space, and free material particles. . . . In judging the application of these auxiliary ideas, we should only demand inner consistency, in which connection special regard has to be paid to the possibilities of definition and observation.[73]

Thus Bohr poses several criteria for the progress of science: on the one hand, we are restricted to descriptive terms drawn from classical science

[72] *APHK*, p. 87; for other remarks on 'abstraction', see *ATDN*, p. 77, and *APHK*, p. 88.

[73] *ATDN*, p. 77.

and to the unambiguous application of such terms in the context of specific observations; on the other hand, however, we may employ such terms as abstractions in describing the 'internal properties' of quantum objects, provided that such usage avoids any inconsistencies.

In other words, the secret of success in scientific method is neither found solely in a positivistic verification through supporting experiments, nor in a Popperian falsification through counter-example. If suggestions like these come into play, it should be noted that both arise out of the assumptions of classical physics which, with its covert emphasis on detached experimentation and strong objectivity, presupposes a grasping of the manifold of experienced reality through theoretical ordering. In such circumstances, the notions of experimental verification or falsification have found a place. They could never be exclusively welcomed by Bohr, however, given his insistence on the impossibility of strong objectivity and the mutuality of observer and observed. As well as verification and falsification, accepting the necessity of our use of 'abstractions', the progress of science will depend on the consistency of our intuitions into the nature of physical reality. Bohr allows for revision and correction in science, as has been stressed, yet not solely as a result of experiment, or solely in terms of the inner consistency of theory. Bohr's notion of scientific knowledge of reality is, it seems, more complex than that. But why?

Perhaps Bohr's account of method is cryptic because it arises less out of a consideration of how science ought to work and more out of a reflection on how he himself came to advance the progress of science. Bohr knew that he 'did physics' successfully, and he knew that his method rested both on relentless analysis and a powerful intuition for 'a little bit of reality'. Heisenberg is by no means the only one to remark on Bohr's great gifts of intuition and the secondary, if necessary, place he gave to mathematics and experiment. Richard Courant has remarked: 'I only want to say, as Harald [Bohr] used to explain, that Niels possessed superhuman intuitive insight into the secrets of nature and that he could perceive the truth without having to translate it into an ordinary human language, including that of mathematics. . . . His intuitive grasp of the situation was overwhelmingly convincing notwithstanding the necessary subsequent farther-reaching quantum-mechanical analysis.'[74] Similarly, Kalckar speaks of Bohr's 'infallible intuition which could immediately grasp the core of a problem in physics'.[75] And

[74] R. Courant, 'Fifty Years of Friendship', in Rozental, *Niels Bohr*, p. 304.

[75] J. Kalckar, 'Niels Bohr and his Youngest Disciples', in Rozental, *Niels Bohr*, p. 230.

Bohr himself used to say: 'You cannot test the soundness of a project unless you go into it with heart and soul.'[76]

A curious characteristic of Bohr's way of doing science was that he could only function, it seemed, in dialogue with his colleagues and students. It was essential for him to spend at least a portion of his time thinking out loud with others. Perhaps he realized that it was in this way that his otherwise subjective intuition fulfilled his own criteria for objectivity.

This may not leave us with a very tidy account of Bohr's view of scientific method, but it mentions all the elements which Bohr regarded as necessary and the conditions under which such elements had proper place. Experiment and theory were equally important, and analysis and synthesis must always be held in balance, because physicists remained party to the descriptions of the processes of nature which they sought to elaborate. The acquisition of knowledge was a possibility and a fact for Bohr. How it occurred remained something of a mystery to him. His great reverence for both nature and knowledge, however, guarded him against false ways of proceeding.

In Mach's account of scientific method, noted in section 1.4. above, theory is reduced to condensed and elaborate reports of observations. A thoroughgoing holist like Rorty, on the other hand, would be sceptical about the objectivity of scientific theories. It is clear from the above discussion that Bohr does not hold with either of these positions and yet endorses aspects of each of them. Holton's assessment of Bohr as the 'best example of a major spirit of science in our century refusing to favour one polar position or the other', is certainly substantiated. But Bohr's moderate holism, as I have termed it, remains elusive and in need of some further clarification.

5.6. *Bohr's Moderate Holism*

So far in this chapter we have concentrated on what appear to be the manifest components of Bohr's philosophy of science. This has entailed a great deal of detective work, of putting clues together, and sketching out distinguishing features. Many questions relevant to a philosophy of science remain unanswered: what does Bohr have to say, for example, about the difference between description and explanation, or about the role of models, or about prediction, dispositions, and powers? His

[76] Quoted by Rozental, ibid., p. 176.

response to these questions can perhaps be inferred, but it is not firmly stated.

Another way of identifying Bohr's view of science involves comparison with rival positions. In this section, therefore, I want to look again at Bohr's position among positivist and non-positivist philosophies of science.

The context for the discussion of recent philosophies of science was given in the introductory remarks to the first chapter, especially in section 1.4. The chief point to be made here is that Bohr does not belong in the camp of the thoroughgoing holists or among the base-grade positivists, and certainly not in the company of those naïve realists who claim a dualist correspondence between scientific theories and independent physical reality. And yet, on the other hand, he embraces aspects of each of these positions as well as occupying ground between them. Given that Bohr stands outside the academic philosophy of science, can his position be connected with theirs? And, saying so much, can he be saying anything at all? Or is he guilty of inconsistency and mystery-mongering? If one adheres to a rigid view of scientific knowing, then there is indeed something suspicious about Bohr's elusiveness; but, given a dynamic and performative account of our participation in nature, Bohr's vision gains credibility.

In his attack on the possibility of the mind mirroring nature, and hence on privileged representations of reality and foundational epistemologies, Richard Rorty provides an alternative, thoroughly holist account of scientific knowledge: scientific fact is reduced to scientific discourse consisting of 'patterns adopted for various historical reasons and as the achievement of objective truth, where "objective truth" is no more and no less than the best idea we currently have about how to explain what is going on'.[77] The key difference between this view of science and Bohr's position is that Bohr sees scientific discourse, through revision and openness, asymptotically converging with the way the world is. Rorty is not interested in any such convergence.

Bohr's new formulation of the objectivity of physics rests on the belief that at least some of our descriptive referring terms work and that nature is real. Rorty does not claim any such factor: one conversation may replace another, but this is not to say that the new conversation is any closer to the truth. Truth, for Rorty, is to be found solely in the consistency of the discourse and not at all in a correspondence between word and world. In contrast to this, while Bohr accepts the circularity

[77] R. Rorty, *Philosophy and the Mirror of Nature* (Oxford, Blackwell, 1980), 385.

of word and world, he also believes that we do advance in our pursuit of nature, although we can never arrive at our ultimate goal. We remain, in the end, part of the nature that we seek to grasp.

I am suggesting that, implicit in Bohr's vision, there is the belief that somewhere along the line the theory is anchored in reality. The success of ordinary language in providing descriptive concepts to match the regularities of the everyday world is the primary source of this belief. In this sense, his philosophy of science displays both realism and moderate holism. Rorty would query both positions, just as he baulks at the possibility of discovering constants in human reflection through the use of transcendental arguments. Opposing both Kant and Habermas in their search for constitutive or indispensable conditions in human knowing, Rorty argues that 'these "subjective conditions" are in no sense "inevitable" ones discoverable by "reflection upon the logic of inquiry." They are just the facts about what a given society, or profession, or other group, takes to be good ground for assertions of a certain sort.'[78] It is not necessary here to say whether Rorty is right or wrong in taking such a stance. My purpose is only to point out that Rorty and Bohr take up differing positions. For it is certain that Bohr, unlike Rorty, regularly appeals to a fundamental epistemological lesson which is universal and necessary and which is based on reflection on the necessary conditions for unambiguous communication.

Bohr's position does entail an assumption 'that we are not dreaming all this', but thereafter it rests on indispensability claims which arise out of reflection on human capabilities. Rorty is interested in clarifying what we can say clearly. Bohr, on the other hand, is interested both in what can be clearly said and also in the truth which dwells in the deep. Bohr is not a thoroughgoing holist, nor can he be described as an out-and-out relativist.

Let us return briefly to the question of Bohr's positivism. Mach's emphasis, it will be remembered, was on the tolerable security of our environment and on the necessary control of truth through appeal to experience. While Bohr embraces aspects of this position, his stress on the balancing of analysis and synthesis shows that, for him, the intelligibility of the theory counts as much in science as the data of experience. For this reason Bohr could 'readily agree with the positivists about the things they want, but not about the things they reject'.[79] If Bohr is a

[78] Ibid., p. 385.

[79] Bohr quoted by Heisenberg in *Physics and Beyond*, pp. 208 f.; see section 1.4. above for the full context of this remark.

positivist, then he does not give the usual unwavering deference to experience which characterizes that stance.

We can also compare Bohr's views with more recent elaborations of positivism. If it is the case that phenomenalism can be condensed into the claim that the way things *are* can be reduced to the way they *appear*, or that our descriptions of objects can be reduced to talk about sense-experience, then this is not Bohr's position. Bohr is much more subtle than that. In constantly warning us against taking classical concepts too seriously, he stresses that our everyday conceptualization of experience may not apply to quantum events: *the way things are requires complementary accounts of the way they seem*. It becomes particularly clear that Bohr is not a phenomenalist, however, when one recalls his willingness to combine 'wave-appearance' and 'particle-appearance' in his description of electrons. No phenomenalist could take such a step in good conscience. The theoretical framework of complementarity is as essential for Bohr as whatever experiential content we may order through it; and complementarity entails at best a very critical phenomenalism, since it insists that one set of experiential data is insufficient for a proper account of experience.

Similarly, Bohr's account of symbols and visualizables runs counter to reductionist versions of positivism. According to this view, theoretical terms have no meaning of their own except in so far as they are reduced to observables. While Bohr agrees that descriptive terms only acquire their meaning from macroscopic circumstances, he also allows for the possibility of using such terms as 'abstractions' and symbols. He would shun the idea of reducing the wave-function in the Schrödinger-equation to a sub-quantum wave. The description remains 'purely symbolic' in a quite different way from the more directly visualizable descriptions of classical physics. Bohr would maintain that quantum theory reflects a facet of nature, namely discontinuity, which could never be observed as such.

Thirdly, according to the instrumentalist refinement of positivism, scientific theories do not either explain or describe reality, but merely offer a means of predicting results. Bohr's probabilist interpretation of quantum calculations indicates a measure of agreement with the instrumentalist position, but he would not accept the irreconcilable separation of theory and reality which instrumentalism implies.

Finally, how does Bohr conform to foundationalist views of science, in which the accurate representation of reality is considered to be a viable goal? In such schemes, as we have seen, a sharp distinction is made between 'inner' mental processes and 'outer' realities. Bohr's reference to visualizables and pictures cannot be interpreted as an appeal to privileged representation of reality. He takes such pictures to be part and parcel of the process of synthesis and communication, and as such they are primary elements in the circle of theory and observation. They are not, however, fundamental essences in an architectonic theory of knowledge.

The context of Bohr's emphasis on the balancing of analysis and synthesis is also more accessible now. The extremes of materialism and mysticism which he wishes to avoid are equally the extremes of a universal mechanist determinism and an anarchic subjective relativism. Bohr's choice, both for a kind of realism and a moderate holism, requires a broadening of the classical assumptions about science and the relationship between theory and observation. The ramifications of this state of affairs have been explored by philosophers of science in terms of the core of a theory, a core which can be regarded as entrenched in reality. It was Bohr's grasp of this core which was the source of his powerful intuition into the nature of physical reality.

Given that Bohr cannot be accurately identified in any of these various guises, he is left in a rather paradoxical position: a holist with the holists, a realist with the realists. This moderate holist, it might be said, wants to have his cake and eat it. It is true that both quantum theory and epistemology were full of paradox for Bohr. His resolution of the dilemma, however, entailed grasping both horns at the same time. It is impossible to grasp each horn with both hands, if one might extend the metaphor, but if one is deft enough it is possible to grasp each horn in either hand: in the everyday world we operate one way, and in the quantum realm we employ another framework. Further, the two points of view are interconnected, for each depends upon the other: though quantum theory offers a more general account than classical physics, it is dependent on classical theory for its terminology.

'Our washing up is just like our language,' Bohr once said, 'We have dirty water and dirty dishcloths, and yet we manage to get the plates and glasses clean.'[80] Put another way, although classical theory is a limited attempt to know of reality, it works perfectly well for us in our everyday lives, just like our language. Complementarity, Bohr's word for moderate holism, allows us to have our cake and eat it, because it prescribes a limit to the application of everyday concepts.

[80] Bohr, quoted by Heisenberg in *Physics and Beyond*, p. 137.

Bohr's contributions to a philosophy of science are so indirect as to be of little value in themselves, but in a wider context the details that we have surveyed here will prove beneficial. We shall now turn to that wider context: first to Bohr's romance with the secrets of nature, and then to our concluding remarks on quantum metaphysics.

6

Mysticism and the Secrets of Nature

I am forcing myself these days with all my strength to familiarize myself with the mysticism of nature.

Niels Bohr

6.1. *Mysticism and Modern Physics*

Bohr's approach to physics cannot, by any stringent standards, be characterized as mystical. For example, William James, perhaps the only 'professional' philosopher with whom Bohr felt any affinity,[1] distinguished four characteristics of mystical experience: ineffability, noetic quality, transiency, and passivity.[2] 'Ineffability' denotes the impossibility of expressing the contents of mystical experience in words; the 'noetic quality' of a mystical state is that it often appears as 'insight into depths of truth unplumbed by discursive intellect', as James puts it; 'transiency' and 'passivity' indicate that such states are relatively brief and that they are experienced almost as if one is in the sway of a superior power. The first, third, and fourth of these characteristics certainly do not fit well with Bohr's practice and philosophy of physics. Indeed, they run counter to his emphasis on communicability and on painstaking analysis. The second characteristic is of interest, however, for the 'noetic quality' of the mystical state corresponds to Bohr's 'almost superhuman intuitive insight into the secrets of nature', as Harald Bohr put it, and it offers a distinctive clue to the understanding of Bohr's mind. It can only be because of this ability that Bohr's vision has been described by some as mystical.

In recent years a number of studies have been made of the mysticism of modern physics. Ken Wilber, for example, in his *Quantum Questions* asks: 'what does it mean that the founders of your modern science . . .

[1] 'I know something about William James,' Bohr said to Kuhn the day before his death, 'I thought he was most wonderful.' *AHQP*, last interview, p. 5 of transcript. See K. M. Meyer-Abich, *Korrespondenz, Individualität und Komplementarität* (Wiesbaden, Steiner, 1965), 133 ff., for comments on Bohr's connections with James.

[2] W. James, *The Varieties of Religious Experience* (London, Fontana, 1960), 367 f.

were, *every one of them, mystics*?'[3] The purpose of this chapter is not so much to add to that branch of popular writing as to set the record straight on Bohr and mysticism. As we shall see, there were frequent references to 'mysticism', most of them sceptical, in conversations between Bohr and his contemporaries. In his essays on the implications of the quantum formalism, Bohr distanced himself from 'a mysticism which is contrary to the spirit of natural science'.[4] Pauli, in particular, wanted Bohr to admit to a kind of mysticism in his way of doing physics, but Bohr strenuously objected to this. By identifying what it was in Bohr's outlook that invited such a challenge, however, we will come closer to grasping his vision of physics and its criteria for success, and, finally, his underlying philosophy.

Bohr himself used the term 'mysticism' both disparagingly and positively. He rejected such a label where it implied subjectivism or high-flown a priori metaphysics, but he praised Newton's near mysticism and used the term to describe his arduous struggles to understand the secrets of nature.

In investigating this particular facet of Bohr's mind, I shall focus on his manner of thinking rather than on his beliefs. I am not claiming that he was a religious mystic in the way that John of the Cross was, for example, nor that he was a nature mystic like Rimbaud. Though I shall touch upon aspects of his attitude towards religion and theology there is no intention of judging Bohr's theism or atheism. My point, rather, is to evaluate the grounds for describing Bohr as a mystic and to identify how his extraordinary powers of intuition shape his view of science and his philosophy of quantum physics.

If Bohr were to be described as a mystic in any way, then the meaning of the term could not be primarily equated with a yearning for ecstatic union with God, nor could it be meant to indicate the transcendence of ordinary thinking and feeling through an immediate apprehension of the unity of everything. It could only be used to capture part of Bohr's emphasis on our unity with nature and the mutuality of spectator and actor. We can discover something about nature not only by experimenting with it, but also by contemplating it: analysis must be balanced by synthesis. It was in such terms that Bohr praised Newton's physics as almost mystical.

[3] K. Wilber (ed.), *Quantum Questions: Mystical Writings of the World's Greatest Physicists* (Boulder, New Science Library, 1984), p. x. See also works like F. Capra, *The Tao of Physics* (London, Fontana, 1976) and G. Zukav, *The Dancing Wu Li Masters* (London, Hutchinson, 1980).

[4] *ATDN*, p. 15; see also p. 116.

Bohr's own colleagues have also sometimes described him in such terms. Oskar Klein, attempting to fathom Bohr's great depth of vision, comments on his efforts to harmonize conflicting realms of experience and logic. In so doing, Klein distinguishes between shallow and profound mysticism:

> While most people tend to notice the differences between similar things, it was natural for him [Bohr] to see what was common to apparently different ones. With his strong feeling for truth, his early concern with the problem of will helped him to be on his guard against cheap generalizations, and to be aware of features of irrationality, arising from the very nature of investigation, in any non-trivial problem. This trend of his had nothing in common with the kind of mysticism, which fills the holes in our attempts towards a rational philosophy with mythological ideas taken literally . . . but it may well be called religious, when that word is used in its essential meaning.[5]

It is this very marked characteristic of Bohr's way of thinking which is the focus of attention in this chapter.

In the following sections Bohr's attitude to religion will be outlined, the charges of mysticism that were laid against him will be discussed, and, finally, some thoughts will be gathered together on the significance of the unveiling of the secrets of nature in Bohr's view of physics.

6.2. *Bohr and Religion*

It is proper, at the outset, to place Bohr's attitude to religious belief in its context. Léon Rosenfeld, colleague and biographer, and not favourably disposed towards religious belief, describes Bohr's position in measured terms:

> He also soon came to share the negative attitude of the progressive bourgeoisie to which his family belonged towards the church and religious beliefs in general; but it is characteristic of his candour and independence of judgement that he only arrived at this conclusion after he had convinced himself that the church upheld doctrines that were logically untenable and shunned the pressing task . . . of alleviating a still widespread pauperism. He never found any occasion in later life to depart from the position of the free-thinker, which he maintained with tolerance and humanity.[6]

More often than not Bohr used the word 'providence' rather than 'God' in his remarks on religious belief. He found the idea of the personifi-

[5] O. Klein, 'Glimpses of Niels Bohr as Scientist and Thinker', in S. Rozental (ed.), *Niels Bohr: His Life and Work* (Amsterdam, North-Holland 1967), 75.

[6] L. Rosenfeld, 'Biographical Sketch', in *NBCW* 1, pp. xx f.

cation of such 'providence' untenable for the same reasons that he objected to the unqualified application of classical concepts to quantum events: there is a limit to the objectivity of ordinary, everyday concepts, and what is applicable within those limits may not be applicable beyond them.[7]

'In religion we renounce the wish to give words an unequivocal meaning from the outset,' Bohr remarked once to Heisenberg, 'while in science we start with the hope—or, if you like, the illusion—that one day it may be possible just to do that.'[8] Bohr thus spoke of classical physics as an 'idealization' (and once as 'idolization'[9]) in much the same way that a too anthropomorphic description of God might be judged idolatrous.

Bohr's attitude to theology and the possibility of the transcendent is somewhat ambiguous. In the context of a discussion about the truth and falsity of ambiguous statements, Dirac recalls Bohr saying that 'There is a God' and 'There is no God' were equally statements of great wisdom and truth.[10] Pauli, himself influenced by Jung's account of our corporate subconscious experience of reality, pressed Bohr throughout a long correspondence to admit to a kind of mysticism. Jaki, on the other hand, claims that Bohr was 'never sympathetic to anything genuinely transcendental, let alone supernatural'.[11]

Despite the urgings of Weizsäcker over several years, Bohr refused to allow any of his essays to be printed in the theological journal *Kerygma und Dogma*. He wanted to avoid any misunderstanding of his interpretation of quantum theory and, in particular, any abuse of the notion of complementarity.[12] When Bohr contributed an article on science and religion to a Festschrift for his friend Johannes Pedersen, he prepared his contribution with great difficulty, and it was more out of friendship for Pedersen than out of a deep interest in his topic.[13] The eighty pages

[7] See letters to Pauli, 13 Dec. 1953, *BSC* : 30, and to Weizsäcker, 30 Dec. 1953, *BSC* : 33; and the note dated 4 Sept. 1954 on 'Spiritual Truth', *MSS* : 21.

[8] Quoted in W. Heisenberg, *Physics and Beyond* (London, Allen and Unwin, 1971), 136.

[9] The transcript of an address given by Bohr at Rooseveldt University, 4 Feb. 1958, reads: 'Classical theories were idolizations . . .'; see *MSS* : 23. Bohr probably intended to say 'idealizations', but the accident of pronunciation is an intriguing one nevertheless.

[10] P. A. M. Dirac, 'The Versatility of Niels Bohr', in Rozental, *Niels Bohr*, p. 309.

[11] S. L. Jaki, *The Road of Science and the Ways to God* (Edinburgh, Scottish Academic Press, 1978), 203.

[12] See the letter from Bohr to E. Schlink, editor of *Kerygma und Dogma*, 23 Oct. 1956, *BSC* : 33, and Bohr's correspondence with Weizsäcker at the same time.

[13] N. Bohr, 'Physical Science and the Study of Religions', in *Studia Orientalia Joanni Pedersen* (Copenhagen, Munksgaard, 1953), 385-90.

of notes that he made whilst he was preparing this article display a restlessness about the relationship between scientific knowledge and religious belief. But Bohr recognized similar problems for physicists and theologians in the application of language to the extraordinary events of the sub-atomic and the supernatural.

The ambiguity of Bohr's attitude to the relationship between physics and religion is evident in the difference between his remarks to the physicist Nielsen and to the theologian Baillie. To Nielsen he observed: 'In physics we do not talk about God but about what we can know. If we are to speak of God, we must do so in an entirely different manner.'[14] During his Gifford Lectures, however, Bohr is reported to have declared to Baillie: 'I think you theologians should make much more use than you are doing of the principle of complementarity.'[15] The comment suggests that Bohr thought the framework of complementarity could provide a possible means of resolving the difficulties which he saw as besetting any discourse about God. If this is so, then there is not going to be a great deal of difference between the conditions applying to language in quantum physics and in theology.

Bohr takes up this line of investigation in his notes for his article on science and religion:

> Without underrating the deep human problems with which we are here concerned a whole new background for the relationship between scientific research and religious attitude has been created by modern development in physics which has demanded a revision of the presumptions for the unambiguous application of our most elementary concepts . . .
>
> . . . it will be attempted to show the development in our time has forced us into epistemological problems of a kind which recalls common problems of the religions.[16]

In another essay written at this time, on 'Unity of Knowledge', Bohr made his point more explicitly: 'In emphasizing the necessity in unambiguous communication of paying proper attention to the placing of the object-subject separation, modern development of science has, however, created a new basis for the use of such words as knowledge and belief.'[17]

[14] Quoted by J. Rud Nielsen in 'Memories of Niels Bohr', *Physics Today* 16 (1963), 27.

[15] Bohr, quoted by J. Baillie in his *Our Sense of the Presence of God* (London, Oxford University Press, 1962), 217. The wire-recordings of Bohr's Gifford Lectures were never edited for publication. Unfortunately only a few notes remain: see *MSS* : 19.

[16] See notes dated 26 and 27 Aug. 1953, *MSS* : 20.

[17] *APHK*, pp. 80 f.

Both science and religion were, for Bohr, attempts to provide greater order and harmony in our existence in the world: 'the very words "sciendum" and "religio" both mean order'.[18] Religion, though, characteristically embraced the more elusive, non-rational dimensions of human existence:

> It is perhaps difficult to conceive of two more contrasting themes than the endeavour to reach a logical description common to all mankind of our experiences regarding that nature in which we ourselves are parts and the religions with [an] aim of standardizing and harmonizing the emotional attitude towards life. . . .
>
> Nevertheless, the study of the history of science and of the religions reveal common traits of the endeavour/position of man in adjusting to fundamental human problems, and notwithstanding the essential different aims and approach, the knowledge of the development and peculiarities of the religious views offer inspiration as regards the contributions which physical science may hope to give in reaching an attitude of ever larger universality and harmony.[19]

In 1955, in a letter to Pauli, Bohr both acknowledged his interest in emotions and rejected the charge of mysticism: 'Indeed, contrary to what some of our common friends seem to believe of me,' he observed, 'I have always sought scientific inspiration in epistemology rather than in mysticism, and how horrifying it may sound, I am at present endeavouring by exactitude as regards logic to leave room for emotions.'[20]

It was at this time, also, that Bohr was widely applying the principle of complementarity to conjugate pairs like life and matter, love and justice. The logic to which Bohr referred in his letter to Pauli, then, may well have been the logic of complementarity. It seems likely that his final position would have taken account of, rather than excluded, religious belief.

Bohr clearly rejected capricious subjectivism and emotional irrationality. This was the kind of mysticism which he deprecated, especially when it impinged upon the interests of natural science. Commenting on the world-view of the ancient mystery religions, Bohr wrote:

> religion may in earlier days be taken to embrace all knowledge beyond the most elementary necessities from daily life, and even what appears as most phantastic phantasies about creation . . .

[18] Undated fragment, *MSS* : 20.

[19] Note dated 15 June 1953, *MSS* : 20, in Bohr's inventive English.

[20] Letter to Pauli, 2 Mar. 1955, *BSC* : 30.

> . . . despite endeavours of a rational philosophy . . . the great Greek school made essential use of a mysticism which is not too clearly distinguished from a sort of knowledge characteristic for religious-like enlightenment in trance and so-called divine revelation.[21]

The rise of science, of course, coincided with the downfall of mystery religions. The 'God of the gaps' was no longer needed. Classical science contributed to a schism between notions of objective knowledge and subjective belief.

Bohr did not wish to expel important elements in human existence from the court of rational enquiry; and he hinted at the unification of knowledge and belief through the reappraisal of the notions of subjectivity and objectivity. Bohr's openness is most clearly illustrated in his readiness to explore realms beyond our ordinary notions of objectivity. While he is critical of any subjective belief which falsely parades as a kind of objective knowledge, he is equally unwilling to accept a narrow materialism which shelters behind superseded canons of objectivity. These extremes of mysticism and materialism can only be avoided by a balancing of analysis and synthesis.

6.3. *Bohr's 'Fruitful Mysticism'*

Terms like 'mystical' and 'mysticism' occur both in writings by Bohr and in comments made about him. Let us look at the context in which they were used, and, especially, how equivocally the terms are employed.

In 1919, wrestling with the fundamental issues in quantum theory, Bohr declared that he was inclined 'to take the most radical or rather mystical views imaginable'.[22] Six years later, in a letter to Heisenberg on the difficulties of sorting out the coupling problem in atomic physics, Bohr declared: 'I am forcing myself these days with all my strength to familiarize myself with the mysticism of nature [die Mystik der Natur].'[23] The honour (or taint) of mysticism was first bestowed on Bohr by his patron Carl Oseen, who described Bohr's original quantum theory as 'a fruitful mysticism'.[24] Remarks in a similar vein were made by Einstein and Born in their reactions to the Copenhagen spirit: Born

[21] Note dated 3 Aug. 1953, *MSS* : 20.

[22] Draft of a letter to C. Darwin, quoted in M. Klein, 'The First Phase of the Bohr-Einstein Dialogue', *Historical Studies in the Physical Sciences* 2 (1970), 20 f.

[23] Letter to Heisenberg, 18 Apr. 1925, *BSC* : 11, *NBCW* 5, p. 362.

[24] As reported to Bohr in a letter from Kramers, 27 Jan. 1919, *BSC* : 4.

described it as 'very mystical' (sehr mystisch), while Einstein called it a 'tranquilizing philosophy—or religion'.[25]

Labels like these arose either out of the way in which Bohr deemed classical physics to apply in one manner but not in another, or out of his revision of the scope and application of ordinary language, or out of the limitation which he imposed on the applicability of the causal space-time frameworks. Although his physics rested on the fruits of mathematical and experimental analysis, it was the power of his mind which produced the new syntheses. It is in this sense that Bohr might almost self-deprecatingly refer to his own work as a kind of mysticism, but he was uneasy and uncomfortable when others applied the term to his way of thinking.

If it was the standard rational generalizations that he was challenging in physics, his broader view was, in his eyes, equally rational. In a comment on his introduction of 'reciprocity' in place of 'complementarity', he declared to Pauli in 1929 that his 'preference for artificial words is due, not so much to an urge for mysticism, as to the endeavour to avoid this by the help of language itself'. Some twenty-five years later, in another letter to Pauli already quoted, he employed a similar disclaimer: 'I have always sought scientific inspiration in epistemology rather than in mysticism.'[26] Here Bohr is rejecting mysticism in its pejorative sense of capricious and emotional subjectivism, but he is also implicitly indicating that his own reflections in epistemology have so altered the ordinary view of science that, from some angles, it seems rather more like mysticism.

Yet Bohr also employs 'mysticism' in a more positive sense, as a contemplative openness to the paradoxes and deep mysteries of human existence. Evidence for Bohr's approval of such a vision can be found in various places in his work. There is more to the exploration of nature than the hoarding of empirical fact. Bohr himself spoke movingly of the need for science to be open to infinite possibility:

> It is not the recognition of our human limitations but our efforts to investigate the nature of these limitations that marks our time. It would only give us a poor picture of our possibilities if we were to compare our limitation with an insurmountable wall. . . . From a deeper and deeper exploration of our basic outlook greater and greater coherence is understood and thus we come to live

[25] Born to Einstein, 15 July 1925, in *The Born-Einstein Letters*, p. 84, and Einstein to Schrödinger in K. Przibram (ed.), *Letters on Wave Mechanics* (London, Vision, 1968), 31.

[26] Letters to Pauli, 31 July 1929, *BSC* : 14, *NBCW* 6, pp. 195, 488, and 2 Mar. 1955, *BSC* : 30.

under an ever richer impression of an eternal and infinite harmony, although we can only feel the vague presence of this harmony but never really grasp it. At every try, in accordance with its nature, it slips out of our hands. Nothing is firm, every thought—yes every word is only suitable to underline a coherence that in itself can never fully be described but always more deeply studied. These are then the conditions for human thought.[27]

This intimation, as R. C. Zaehner puts it, 'of something which transcends and informs transient Nature', cannot itself be classified as mystical, but it belongs at the edge of nature mysticism.[28]

Bohr, in a similar frame of mind, once described Newton's 'deep occupation' as 'tending almost to mysticism'. He then went on to warn that 'all talk of distinguishing between rationalism and mysticism is essentially ambiguous'.[29] Here he is using the term in a positive sense, indicating powers of contemplation, insight, and synthesis. Indeed, our word 'theory' comes from the Greek verb 'to contemplate'. This understanding of mysticism is therefore quite the reverse of his other remarks that mysticism connotes an escape into vagueness.[30]

In 1950 Pauli sent Bohr a long letter in which he moved on from the discussion of epistemological issues in the interpretation of quantum theory to some reflections on the nature of mysticism.[31] Bohr's enigmatic reply simply conceded that 'they understood each other so well',[32] but this gave Pauli no satisfaction whatsoever. Ten years later, immediately after returning from a visit to Copenhagen, Pauli fired off a more pressing letter at Bohr. He stated that he wished 'to come back to our talk of yesterday on the connection between the concepts "God" and "knowledge" '. He then continued by discussing various attitudes towards God as a way of challenging Bohr's notion of 'providence'. 'I only wish to emphasize', said Pauli admonishingly, 'that one has to *know* all this extra-church-tradition if one discusses such questions as you did yesterday.'[33]

[27] Address given in Copenhagen, 21 Sept. 1928, *MSS* : 11, p. 4.

[28] See R. C. Zaehner, *Mysticism Sacred and Profane* (Oxford, Oxford University Press, 1961), 35; see also pp. 198 f.

[29] See second draft of 'Newton's Principles', *MSS* : 18, p. 27, and note dated 12 Aug. 1946, *MSS* : 17, p. 10.

[30] As evident in the eighth draft of 'The Unity of Human Knowledge', 27 Sept. 1960, *MSS* : 24, p. 7.

[31] Pauli to Bohr, 3 Oct. 1950, *BSC* : 30. This is a nine-page letter in barely decipherable German, but see pp. ia, iia.

[32] Letter to Pauli, 23 Dec. 1950, *BSC* : 30.

[33] Pauli to Bohr, 15 June 1952, *BSC* : 30.

Bohr appeared to misunderstand Pauli's references to the *via negativa* and to the mystic's knowledge of God as the unknowable one. In his letter, Pauli referred to the non-personal Gods of Buddha and Lao Tse, to the interiorly discovered God of Meister Eckhart, and to the God known only through a knowledge of what God is not. Bohr's reply was firm in its insistence on 'the logical difficulties which the perception of a personified providence meets'.[34] Although Bohr doubtlessly found Pauli exasperating on this issue, as on many other issues, Pauli was perhaps rightly aggrieved. He responded by saying that he had 'hoped to hear more' from Bohr,[35] but Bohr had no more to say on the matter at that time.

Privately, however, Bohr continued to work out his ideas more completely. In the tortured sketches for his essay on science and religion, written shortly after the correspondence with Pauli, we find the undeveloped heading, '*Mysticisme og atomteori*'.[36] When Pauli wrote again in 1955, commenting that he did 'find sometimes scientific inspiration in mysticism' and that 'the "Unity" of everything has always been one of the most prominent ideas of the mystics', Bohr did not take the bait. Although he himself was constantly appealing to 'unity', he responded rather shortly to Pauli: 'it is a pure discussional accident which words, like mysticism or logical systematism, the one or the other of us uses'.[37]

There is more than dismissal entailed in this curt reply to Pauli. Bohr also has in mind the need, as he saw it, to balance the methods of analysis and synthesis: 'our only way of avoiding the extremes of materialism and mysticism is the never-ending endeavour to balance analysis and synthesis.'[38] In these last remarks to Pauli, he is appealing more to the continuity between analysis and synthesis than to their disjunction. Again, given that Bohr's 'logical systematism' operates on a broader scale than our usual more narrow notions of rationality allow, he regards as 'logical' the sort of vision which Pauli wishes to describe as 'mystical'. The logic Bohr has in mind, of course, is the seemingly paradoxical logic of complementarity.

A similar state of affairs exists in Bohr's remarks on the continuity of spirit and matter. In the following quotations, from his notes of 1954,

[34] Letter to Pauli, 31 Dec. 1953, *BSC*: 30, p. 4, from the Danish.
[35] Pauli to Bohr, 19 Feb. 1954, *BSC*: 30.
[36] See note dated 25 Aug. 1953, *MSS*: 20.
[37] Pauli to Bohr, 11 Mar. 1955, and Bohr to Pauli, 25 Mar. 1955, *BSC*: 30.
[38] 'Analysis and Synthesis in Science', p. 28.

one finds not only evidence for his vision of the unity or mutuality of spirit and matter, but also of the indwelling of spirit in matter:

> The very problem of a spirit behind existence is certainly undefinable [?] if it shall not merely mean a symbol for an ultimate harmony which according to the very word cannot be analyzed nor capable of objective description.
>
> The question is not about subjective belief but about a serious endeavour at analyzing the situation and definition of the words by which it can be objectively described. . . .
>
> As regards the question of spiritual truth, I shall not repeat what has already been said of the inherent inseparability of materialistic and spiritualistic views. . . .
>
> . . . materialism and spiritualism, which are only defined by concepts taken from each other, are two aspects of the same thing.[39]

It seems to me that Bohr is resolving the perennial dichotomy of matter and spirit with exactly the same strategy that he applied to the opposition of analysis and synthesis or the incompatibility of quantum and classical physics. Each has its proper place provided one respects the conditions that apply.

Though Bohr rejects the possibility of our talking about a purely supernatural 'providence', since such an entity lies beyond all possible experience and perhaps also because of the logical difficulties posed by the problem of evil, he endorses our continued endeavour to grasp the infinite and elusive harmony which lies at the heart of our encounter with nature. This is the secret of nature which his science pursues.

6.4. *Secrets of Success*

Bohr possessed, as his brother Harald observed, 'superhuman intuitive insight into the secrets of nature'. He believed in the existence of 'an ultimate harmony' or 'an external and infinite harmony' as fundamental to the nature of the world in which we found ourselves.[40] And, even though this ultimate harmony remained elusive and incapable of objective description, he believed that science, in so far as it ordered augmented experience, provided 'an attitude of ever larger universality and harmony'.[41] The secret of the success of science lay not just in its

[39] See the notes under the heading 'Spiritual Truth', 1 and 4 Sept. 1954, the typed draft, 21 Aug. 1954, and the note dated 19 Aug. 1954, all in *MSS* : 21.

[40] Address given in Copenhagen, 21 Sept. 1928, p. 4, and notes on 'Spiritual Truth', 1 Sept. 1954, *MSS* : 21. See nn. 26 and 38 above.

[41] Note dated 15 June 1953, *MSS* : 20.

clarity and objectivity, therefore, but in the convergence of its unified frameworks towards the secrets of nature which it explored.

It is at this point that Bohr's rejection of positivism and pragmatism becomes most sharply stated. Heisenberg recalls a conversation between Bohr and Pauli in the following terms. First Pauli remarked:

> The positivists have gathered that quantum mechanics describes atomic phenomena correctly, and so they have no cause for complaint. What else we have had to add—complementarity, inference of probabilities, uncertainty relations, separation of subject and object, etc.—strikes them as just so many embellishments, mere relapses into scientific thought, bits of idle chatter that do not have to be taken seriously. Perhaps this attitude is logically defensible, but, if it is, I for one can no longer tell what we mean when we say we have understood nature.

Heisenberg then went on to report Bohr's response to Pauli, quoted in full in the opening chapter, with Bohr's objections to positivism: 'its prohibition of any discussion of the wider issues simply because we lack clear-cut enough concepts in this realm,' said Bohr, 'does not seem to be very useful to me.'[42] Concluding his recollections of this conversation, Heisenberg remarks: 'if we may no longer speak or even think about the wider connections, we are without a compass and hence in danger of losing our way'.

This is not an isolated example of Bohr's interest in attempting to penetrate the eternal questions. Jørgen Kalckar reminisces on several such dialogues in the following passage:

> Best of all was when the conversation turned to the so-called 'eternal questions'. Nowhere did Bohr's influence as an educator have a more profound significance for his pupils than in our view of and feeling for these and their inseparability from the total conception of nature. . . . These talks, often resumed and continued in one's mind, imparted an overwhelming and unforgettable impression of the unity and inherent harmony in Bohr's attitude to existence and his conception of its richness and diversity.[43]

Bohr's sense of deep truth is tied in with his strongly asserted, if elusive, view of quantum objectivity. Against Rorty he would have argued that science is more than a continuing conversation, for it has its secrets of success. And yet Bohr never fully defended this position to his opponents' satisfaction. This attitude of Bohr's, apparently sitting on

[42] Heisenberg, *Physics and Beyond*, pp. 208 f. See section 1.4. above, at n. 30.

[43] J. Kalckar, 'Niels Bohr and His Youngest Disciples', In Rozental, *Niels Bohr*, p. 235.

the fence between anarchy and order, disturbed Einstein. Bohr later recalled: 'On his side, Einstein mockingly asked us whether we could really believe that the providential authorities took recourse to dice-playing ("... *ob der liebe Gott würfelt*"), to which I replied by pointing at the great caution, already called for by ancient thinkers, in ascribing attributes to Providence in everyday language.'[44] Behind Bohr's jesting lies a deep conviction that his view of physics does not abandon objectivity and, in particular, that quantum physics is concerned with fathoming the profound truths of nature.

In a thoroughly relativist account of knowledge there are neither foundations, nor claims for absolutes. In foundational epistemologies, on the other hand, two extreme positions are possible. First, one can hold something like a Marxist materialism: truth arises from the way the world is and its inherent laws of material and social evolution through history. Or, at the opposite extreme, one can appeal to something akin to Hegel's absolute spirit, to a metaphysical 'ground of being' which provides an ultimate basis for reality, objectivity, and truthfulness. In both schemes a claim is made for the existence of subject-independent entities. The knot of theory and observation is ultimately cut through. We are assured that our conversations are anchored in objective reality, no matter how we may view that reality. So also, in scientific practice, we have a guarantee that we are getting closer to the whole truth and that there is a truth that we can seek.

But Bohr does not give himself solely to either of these positions. Nor, as we have seen, does he agree with Einstein's insistence on a complete, causal account. Bohr, indeed, leaves several philosophical issues unresolved in his appeal to complementarity. His stance has some justification, all the same, if his vision includes a secret of success. In his own terms, though, such a secret necessarily dwells in the deep.

Rorty, defending the paradoxical position of the thoroughgoing holist, is forced to reject any philosophical notion of a secret of success hidden beneath the antics of science and epistemology. In a very enlightening discussion with H. L. Dreyfus and Charles Taylor, Rorty attempts to deal with the apparent success of science in the following terms:

RORTY. But there is no answer to the question, 'How did the scientists manage to do it?' In particular, I don't think it helps to offer as an answer to the latter question, 'Because they found some subject-independent terms.' That's because I don't think *anything* would help.

[44] *APHK*, p. 47.

DREYFUS. . . . I want to attribute to the later Heidegger the view that 'only a God can save us.' As I understand it, this refers to a very particular problem, namely the problem of finding a new paradigm . . . that can focus our dispersed inherited micropractices and linguistic practices. I think that Heidegger is right; there's nothing that we can do to find this new focus . . .

RORTY. I would want to disjoin the notion that a focus and a paradigm is needed from the notion that it is the task of philosophy to give it to us.

DREYFUS. No, that's the way God comes in; it is not the task of philosophy.

RORTY. Yes, but there ought to be a middle ground between philosophy professors and God.

DREYFUS. Who, for instance . . . ?

TAYLOR. . . . Rorty seems to be holding that his is not a question, that nothing can be said, that our natural sciences work but that nothing can be said about why they work. What can we converse about if we encounter ineffability everywhere?

RORTY. It seems to me that we don't encounter ineffability if we think that a question probably doesn't have a usable answer. . . . Kuhn's reduction of philosophy of science to sociology of science doesn't point to an *ineffable* secret of success; it leaves us without the notion of the secret of success.[45]

In the last comments Rorty makes it clear that his kind of holism cannot include any notion, stated or otherwise, of the possibility of a secret of success. Scientific discourse elaborates on scientific discourse rather than on nature.

Another way of taking Rorty's point is to see it as a defence of Wittgenstein's 'therapeutic' view of philosophy: he wants to teach us how to prevent ourselves from getting headaches by pointing out that there are barriers to our knowing which we ought to stop banging our heads against. Philosophy can take us no further in these directions. But, and here light is cast on Bohr's position once again, there is another side to Wittgenstein's vision. While we should remain silent about things of which we cannot speak, that is not to deny the mystical. As the younger Wittgenstein put it: 'There are, indeed, things that cannot be put into words. *They make themselves manifest*. They are what is mystical.'[46] Characteristic of Bohr's complementarity is the wish to take up both positions: on the one hand our language is restricted by the scope of our everyday images and frameworks; and on the other hand our every act of objectification also makes manifest, to use Wittgenstein's terms, the elusiveness of the subject.

[45] H. L. Dreyfus, R. Rorty, and C. Taylor, 'A Discussion [of Holism and Hermeneutics]', *Review of Metaphysics* 34 (1980), 53 ff.

[46] See Wittgenstein's *Tractatus Logico-Philosophicus*, 6.522, 6.44, and 6.45.

The 'making manifest', indeed, has about it the nature of a disclosure, a discovery of that which is constitutive as a necessary condition. The mystic's intimation of the infinite through contemplation may not be the same as the transcendental philosopher's insight into the necessary conditions constituting our contingent activity, but the two movements are similar enough to be mistaken for each other.

Similarities between Bohr and Einstein can also be found here. Though the two may have differed about the future of physics, with Einstein seeking more stringent explanation, their visions of the harmony of nature were much more alike. Einstein was quite insistent about his own kind of theism, which in itself had to do with the secrets of success of science: with his refined scientific nose, he was constantly seeking out the scent of 'the Old One'. Bohr, less definite perhaps, wondered at the infinite, elusive harmony of nature.

In 1939 Einstein addressed the Princeton Theological Seminary thus:

> But science can only be created by those who are thoroughly imbued with the aspiration toward truth and understanding. This source of feeling, however, springs from the sphere of religion. To this there also belongs the faith in the possibility that the regularities valid for the world of existence are rational, that is, comprehensible to reason. I cannot conceive of a genuine scientist without this profound faith. The situation may be expressed by an image: science without religion is lame, religion without science is blind.[47]

In a letter to Maurice Solovine in 1952 Einstein makes it clear that his unorthodox belief arises not as the result of some a priori argument, but rather as made manifest through wonder:

> You find it remarkable that the comprehensibility of the world (insofar as we are justified to speak of such a comprehensibility) seems to me a wonder or eternal secret. Now, *a priori*, one should, after all, expect a chaotic world that is in no way graspable through thinking. One could (even *should*) expect that the world turns out to be lawful only insofar as we make an ordering intervention. It would be a kind of ordering like putting into alphabetic order the words of a language. On the other hand, the kind of order which, for example, was created through [the discovery of] Newton's theory of gravitation is of a quite different character. Even if the axioms of the theory are put forward by human agents, the success of such an enterprise does suppose a high degree of order in the objective world, which one had no justification whatever to accept *a priori*. Here lies the sense of 'wonder' which increases ever more with the development of our knowledge.

[47] A. Einstein, *Ideas and Opinions* (New York, Crown, 1954), 46.

And here lies the weak point for the positivists and the professional atheists, who are feeling happy through the consciousness of having successfully made the world not only god-free, but even 'wonder-free'. The nice thing is that we must be content with the acknowledgement of the 'wonder', without there being a legitimate way beyond it. I feel I must add this explicitly, so that you wouldn't think that I—weakened by age—have become a victim of the clergy.[48]

For all their differences, the sharing of this profound depth and breadth of vision was surely an important factor in the deep affection and admiration which Bohr and Einstein had for each other. Einstein, certainly, declared his confidence in Bohr's way of thinking: 'like one perpetually groping and never like one who believes to be in possession of definite truth'.[49]

The conscientious philosopher may be frustrated by this hand-waving at the secret of success of science to which Einstein clearly confesses and which has been uncovered in Bohr's writings. But at least if the case made here is correct, we now have a better idea of where Bohr stands. His views of science, of reality and objectivity and truthfulness, are framed not just by philosophical arguments of a rigorous logical kind, but also by an underlying belief that 'we are not dreaming all this'. Having said that, in Bohr's own terms it must also be declared that the extremes of mysticism and rationalism remain repudiated, as do the extremes of subjectivism and materialism.

Edward MacKinnon commented both perceptively and sympathetically on the manner in which Einstein forges his 'theological escape route':

An epistemological position that relies on a natural affinity between human rationality modeled on a deductive system and immanent cosmic rationality couched in pantheistic terms will inevitably be unacceptable to most for a variety of scientific, epistemological, and theological reasons. It must, however, be admitted that those of us who reject it do not have to live with the miracle of Einstein's creativity. He did. This miracle . . . drove him into a position peculiarly reminiscent of Newton's feeling of a special affinity to God and of a willingness to invoke God to close the gaps in his system.[50]

However, MacKinnon then moves on to contrast Einstein's position with what he calls Bohr's 'more coherent, but more limited, account'.

[48] Letter to Solovine, 30 Mar. 1952, quoted in A. P. French (ed.), *Einstein: A Centenary Volume* (London, Heinemann, 1979), 162 f.

[49] See A. Pais, *'Subtle is the Lord . . .': The Science and Life of Albert Einstein* (Oxford, Oxford University Press, 1982), 417, and the beginning of Chapter 1 above.

[50] E. MacKinnon, *Scientific Explanation and Atomic Physics* (Chicago, University of Chicago Press, 1982), 389.

He takes Bohr's focus on language and meaning to be narrower than Einstein's vision of the ultimate vindication of theory. This is indeed true, but this chapter has shown that Bohr's position entailed a serious acknowledgement of those profound truths which defy articulation and yet which provide foundations for all our more familiar activities.

If Bohr restrains himself from talking too much about the mystical and about ultimate harmony, it is because his epistemology will not permit him to do so. Any interest which Bohr showed in 'religion, orthodoxy and mysticism', Kalckar remarks, 'could be directed solely to the investigation of their psychological and social roots'.[51] Such an observation is important and serves as a stern caution against making too much of Bohr's remarks on religion and mysticism. On the other hand, the same epistemology, based as it is on the limitations to human experience and language, manifests an openness to unlimited and ultimate harmony. Bohr believed that, through continued revision and ever-increasing precision and comprehension, science brought us closer and closer to understanding that nature of which we are a part. Such a goal of ultimate unity may never be reached scientifically, since it defies analysis and synthesis. The ideal is not abandoned, however, but is implicitly retained as proleptic justification of scientific activity.

I have not attempted to foist on Bohr a religious position which he himself would not have wanted to adopt. Rather, I have tried to draw out some implications of his position and to support these with evidence from his own writings. It is perhaps 'a pure discussional accident' whether we call his position mystical or logical, as Bohr declared to Pauli. By his own admission, however, Bohr's logic goes beyond the usual range of classical scientific discourse, and Pauli was justified in pressing Bohr to see the consequences of his stance from another angle.

It is arguable, therefore, that Bohr worked with a secret of success in mind. He called it 'die Mystik der Natur'. If this is correct, then it is also a further indication that Bohr's notion of complementarity entails a moderate and not a thoroughgoing holism. One could never substantiate the claim that Bohr ever enjoyed mystical experience, nor would I want to do that. There is considerable support none the less for Bohr's acceptance of 'an ever richer impression of an eternal and infinite harmony' and 'a coherence that in itself can never be fully described but always more deeply studied'. This belief in a One in the Many is not just the secret of success of science as Bohr views it, but also a clue to a rational metaphysics. This is a trace of universalism inherent in Bohr's

[51] Kalckar, 'Niels Bohr and his Youngest Disciples', p. 236.

point of view. Although it does not draw him beyond the bounds of empiricism and language, since such bounds constitute our effective communication and our sense of reality, the universal harmony remains an acknowledged factor in his physics. So, finally, we come to examine Bohr's thinking in terms of the overlap between metaphysics and microphysics.

7

Microphysics and Metaphysics

> I could see no reason why the prefix 'meta' should be reserved for logic and mathematics . . . and why it was anathema in physics.
>
> Niels Bohr

7.1. *Descriptive Metaphysics*

Different aspects of Bohr's way of thinking have been considered in turn in the previous chapters. Having set this analysis in motion, it is now proper to balance the taking apart with some putting together, some synthesis. This concluding chapter introduces the notion of 'descriptive metaphysics' and discusses the metaphysical character of Bohr's philosophy. Focus is centred in particular on the three fundamental propositions, identified at the end of Chapter 3, on which Bohr's interpretation rests.

In the first chapter the testimonies of Einstein, Heisenberg, Jordan, and de Broglie were presented: Bohr was a seeker after deep truths, primarily a philosopher, one who transformed our ways of looking at reality, the Rembrandt of physics with a predilection for 'obscure clarity'. Attention was then concentrated on Bohr's interpretation of the quantum formalism and his transcendental philosophy: the prescription of complementarity was seen to rest on three fundamental arguments listed as:

(B1) Some kind of conceptual framework is a necessary condition of the possibility of ordering experience.

(B2) It is a necessary condition of the possibility of objective description of processes at the boundaries of human experience that concepts related to more normal experience be employed.

(B3) Our position as observers in a domain of experience where unambiguous application of concepts depends essentially on the conditions of observation demands the use of complementary descriptions if the description is to be exhaustive.

In Chapter 4 the unfolding of these arguments in the debates with Einstein was reviewed, and this in turn led to a presentation of Bohr's philosophy of physics, with particular emphasis on his attitudes towards realism, objectivity, the role of symbols and pictures, and the balancing of analysis and synthesis. Finally, in Chapter 6, evidence was given of Bohr's underlying belief in the infinite harmony of nature, and thus for his sense of a secret of success for science.

At the beginning of this work it was noted that Bohr disliked being labelled. Despite this, I have tested out several identifications: transcendentalist, moderate holist, and relative realist. By way of synthesis, now, I want to establish some connections between Bohr's interests and those of the eminent descriptive metaphysicians. His three key arguments, listed above, certainly provide an avenue along which to approach the discussion of fundamental questions about experience, language, and knowledge.

The term 'descriptive metaphysics' was fixed into philosophical discourse by Strawson: 'Descriptive metaphysics is content to describe the actual structure of our thought about the world, revisionary metaphysics is concerned to produce a better structure.'[1] In drawing on this distinction, the purpose, once again, is not to constrain Bohr's spirit, but rather to indicate where it might be placed in the sweep of human thought so that our grasp of Bohr's vision might be more secure. Bohr was not one to offer a revision of the theory of knowledge, but he did attempt to clarify some of the fundamental conditions applying to our communication of our experience of the world.

Against this view it might be argued that Bohr's appeals to an epistemological lesson and his introduction of complementarity amount to producing a better structure of our thought about the world. But I would argue that Bohr's 'rational generalization' is much more a revision of our descriptions of nature than a restructuring of our thought about how nature is. Whereas revisionary metaphysicians like Leibniz and Spinoza offer a new ontology of being, the descriptive metaphysicians are more concerned with our interaction with the world. It is interesting here to recall Bohr's irreconcilable differences with Einstein: where Bohr was preoccupied with epistemology, Einstein tended more to ontology.

It should be noted at the outset that Bohr explicitly dissociated himself from what he saw as the empty grandeur of lofty metaphysics. On one occasion, after giving his usual account of the subject-object paradox and the conditions for unambiguous communication, he then

[1] P. F. Strawson, *Individuals* (London, Methuen, 1959), 9.

remarked that there was 'no question whatever . . . of purely metaphysical speculation'.[2] The qualifier 'purely' is deliberately placed.

It is true that Bohr could hardly be described as engaging in the pure a priori metaphysics of, say, Descartes, Spinoza, or Wolff. These more abstract and ambitious ways of answering all human questions about ultimate reality and certain knowledge have variously been described as 'transcendent', 'reductive-deductive', and 'revisionary'.[3] These more speculative philosophers have reduced metaphysical questions to a handful of premisses and then deduced and reconstructed an almost visionary account of the unity (or duality) of knowing and being. The abiding difficulty with revisionary metaphysics, no matter how enlightening its conclusions, is that the controlling premisses are based more in speculation than in experience and hence are always open to question. The emerging results are thus equally suspect. Such metaphysicians, moreover, often dismiss the data of sense-experience as illusory and inadequate for the kind of certitudes they seek. In abdicating from the court of experience they find themselves just as far removed from the world of science. This revisionary metaphysics is, surely, the 'purely metaphysical speculation' which Bohr disowns. He would rather follow experience than a priori arguments.

However, there is a moderate metaphysical alternative to rejecting metaphysics out of hand. Here the starting-point is our experience of the world and all that such experience can be seen to entail: our knowledge and our ignorance, our language and our silence, our activity and our passivity. Variously denoted as 'immanent', 'inductive', and 'descriptive', such a metaphysics can also be described as 'transcendental': it seeks to articulate that which lies submerged in the actual structure of our participation in the world. As Strawson describes his own project in descriptive metaphysics, it aims to lay bare the most general features of our conceptual structure:

> For there is a massive central core of human thinking which has no history—or none recorded in the histories of thought; there are categories and concepts which, in their most fundamental character, change not at all. Obviously these are not the specialities of the most refined thinking; and are yet the indispensable core of the conceptual equipment of the most sophisticated human beings. It is with these, their interconnexions, and the structure that they form, that a descriptive metaphysics will be primarily concerned.[4]

[2] Bohr, 'Causality and Complementarity', p. 296.

[3] See W. H. Walsh, *Metaphysics* (London, Hutchinson, 1963), 84; F. C. Copleston, *Religion and the One: Philosophies East and West* (London, Search, 1982), 4-8; and Strawson, *Individuals*, p. 9.

[4] Strawson, *Individuals*, pp. 9 f.

Such an approach is also characterized by a cautious realism and, given its transcendental method, a moderate holism.[5] It is not far removed from Bohr's considerations in his interpretation of quantum theory. And, if Bohr can be compared with the prototypical descriptive metaphysicians, Aristotle and Kant, then their work may also shed some light on obscurities in his own endeavours. Bohr's uniqueness, however, would make such a task extremely demanding.

Bohr has indeed spoken in favour of a moderate kind of metaphysics. In a conversation reported by Heisenberg he expressed his view in something like the following terms:

> Philipp Frank . . . gave a lecture in which he used the term 'metaphysics' simply as a swearword or, at best, as a euphemism for unscientific thought. After he had finished, I had to explain my own position, and this I did roughly as follows:
>
> I began by pointing out that I could see no reason why the prefix 'meta' should be reserved for logic and mathematics—Frank had spoken of metalogic and metamathematics—and why it was anathema in physics. The prefix, after all, merely suggests that we are asking further questions, i.e., questions bearing on the fundamental concepts of a particular discipline, and why ever should we not be able to ask such questions in physics? . . . In all such discussions what matters most to me is that we do not simply talk the 'deeps in which the truth dwells' out of existence. That would mean taking a very superficial view.[6]

Bohr's interest in 'the fundamental concepts of a particular discipline' matches Strawson's exploration of 'the actual structure of our thought about the world'. Bohr's fundamental propositions in his interpretation of the quantum formalism, statements (B1) to (B3) above, are precisely the results of such an investigation.

'Metaphysics' is of course a very open term: it was the accidental title given to Aristotle's lectures on the ultimate principles and causes in human knowing. Quite fortuitously, the works that came after the *Physics—ta meta ta physika*—are concerned with questions which arise as implications of our ability to do physics. How are we able to know beyond ordinary experience? What can we possibly know beyond ordinary experience? How can we affirm a Many of particular things without acknowledging the being of an underlying One? If there is a One, how can there be Many? The battlefield of these endless controversies, as Kant put it at the beginning of the *Critique of Pure Reason*, is called metaphysics.

[5] See, for example, Strawson's comments in *Individuals*, pp. 40, 198.

[6] Quoted by Heisenberg in *Physics and Beyond* (London, Allen and Unwin, 1971), 210.

If they were able to finish off the battle, metaphysicians would then be able to take the high ground of certain knowledge and win the war against our scepticism. Metaphysics was traditionally distinguished from natural science as being more synthetic than analytic, more theoretical than practical, more speculative than conclusive. In the aftermath of the rise of classical science, it fell from favour along with witchcraft and divination and became less than fashionable. Natural science, dealing with immediate experience and achieving one success after another, upstaged the conversations of philosophers and theologians. Questions of a metaphysical character nevertheless remained; science, after all, is not self-explanatory, even though one could choose to dismiss queries about its nature as meaningless or unanswerable.

Bohr's physics went far beyond experimental and mathematical analysis, although such activities were essential to science. His focus on the nature of our concepts may even have anticipated and shaped his physics. He was as interested in trying to clarify that which was unclear as he was in stating that which was obvious. His notion of complementarity is not so much a revision of metaphysical questions, however, as an attempt to identify, in the style of descriptive metaphysics, what can be said and what cannot be said. And, as argued in Chapter 3, this notion arises as much out of epistemological considerations as out of the implications of quantum theory. Of relevance here is the story of Bohr and two colleagues engaged in conversation, on a sailing trip, around the time of his formulation of the principle of complementarity:

> Bohr was full of the new interpretation of quantum theory, and . . . there was plenty of time to tell of this scientific event and to reflect philosophically on the nature of atomic theory. Bohr began by talking of the difficulties of language, of the limitations of all our means of expressing ourselves, which one had to take into account from the very beginning if one wants to practice science . . . Finally, one of the friends remarked drily, 'But, Niels, this is not really new, you said exactly the same ten years ago'.[7]

Some would say that Bohr stresses the epistemological to the detriment of the ontological. Shimony states: 'Bohr's thought can be related to a certain philosophical tradition (including Hume and Kant) in which a theory of knowlege is worked out without commitment to a theory of existence.'[8] Perhaps one reason for this approach is that, although he

[7] See W. Heisenberg, 'Quantum Theory and its Interpretation', in S. Rozental (ed.), *Niels Bohr: His Life and Work* (Amsterdam, North-Holland, 1967), 106 f.

[8] A. Shimony, 'Physical and Philosophical Issues in the Bohr-Einstein Debate', Proceedings of the Bohr Centenary Symposium of the American Academy of Arts and Sciences (forthcoming), p. 3 of preprint.

wants to shape a realist metaphysics, he wants equally to avoid the usual less critical realism implicit in classical physics. The combination of classical concepts and classical frameworks does not adequately describe reality, especially in its fine structure. The secrets of nature are approached asymptotically rather than captured directly in representational concepts. Though he does not speak much of the power of intuition in coming to grasp the nature of reality, Bohr's own way of doing science, as well as his emphasis on the need to balance analysis with synthesis, indicates that an exploration of our ways of knowing is as important in a realist metaphysics as a consideration of the reality that we do experience. For Bohr, the important question was not the existence of reality or nature, but the proper description of nature. Because of this, epistemology takes priority over ontology, and speculation about a metaphysics of nature is dismissed as misleading. In using his transcendental reflections on the constitutive conditions for observation, he shifted between the manifold of experience to the indispensable underlying constraints. This may not be a movement from the Many to the One, but it is a movement between the two. Descriptive metaphysics does not take us very far, proceeding cautiously rather than boldly, but it guides us in what we can and cannot say.

Descriptive metaphysics is situated somewhere between revisionary metaphysics on the one hand, and positivism and relativism on the other. In seeking to explore that which lies submerged beneath the structures of our familiarity with the world, it takes the given as both multiple and whole. Transcendental arguments, claiming to uncover that which is constant in our abilities and activities, do have a circular character about them. Critics will term them 'parasitic', while defenders of such methods will describe them as 'self-referential' or 'performative'.[9] The descriptive approach will always entail a kind of holism or complementarity, but it attempts to go beyond hapless circularity. While it may not answer all our questions, neither does it leave us zombie-like in a cave of random shadows.

So far, I have given an outline of metaphysics, and of descriptive metaphysics in particular. My point has been to indicate some of the interests and methods which Bohr has in common with more recognized

[9] See R. Rorty, 'Verificationism and Transcendental Arguments', *Nous* 5 (1971), 4; T. E. Wilkerson, 'Transcendental Arguments Revisited', *Kant-Studien* 66 (1975), 114; and C. Taylor, 'The Validity of Transcendental Arguments', *Proceedings of the Aristotelian Society* 79 (1978-9), 151-65.

philosophers, as well as to touch on the connections between transcendental argument and moderate holism. Before examining Bohr's three key arguments in detail, I have been indicating the sort of company in which, as a philosopher, he might be comfortably situated.

Strawson introduces his own modest essay in descriptive metaphysics with an observation which is relevant here:

> though the central subject-matter of descriptive metaphysics does not change, the critical and analytical idiom of philosophy changes constantly. Permanent relationships are described in an impermanent idiom, which reflects both the age's climate of thought and the individual philosopher's personal style of thinking. No philosopher understands his predecessors until he has re-thought their thought in his own contemporary terms; and it is characteristic of the very greatest philosophers, like Kant and Aristotle, that they, more than any others, repay this effort of re-thinking.[10]

Bohr's metaphysics is couched in the framework of microphysics. He either lacked the inclination or the wherewithal to rethink the contributions of his predecessors. His may not have been a purely metaphysical speculation, but it had a great deal to do with the necessary conditions for the possibility of doing quantum physics.

Wishing to avoid the extreme of a purely a priori revisionary metaphysics, Bohr was aware of doing a more modest kind of metaphysics in his own work. In 1929, whilst preparing the article for the Planck Jubilee, he wrote to R. H. Fowler: 'I am trying in these days to finish a small note of the philosophical aspects of quantum theory which I hope will not be considered to be too metaphysical.'[11] We shall now re-examine his fundamental arguments in the context of such modest or descriptive metaphysics.

7.2. *The Fundamental Arguments*

'Far from containing any mysticism foreign to the spirit of science,' says Bohr, 'the notion of complementarity points to the logical conditions for description and comprehension of experience in atomic physics.'[12] Bohr's lament to Kuhn on the eve of his death: 'I think it would be reasonable to say that no man who is called a philosopher really understands what one means by the complementarity description. . . . They did not see that it was an objective description, and that it was the

[10] Strawson, *Individuals*, pp. 10 f.

[11] Letter to R. H. Fowler, 14 Feb. 1929, quoted in *NBCW* 9, p. 556.

[12] *APHK*, p. 91.

only possible objective description'[13] is reminiscent of Kant's despair at the early interpretation of his work as lofty metaphysical idealism. Bohr regarded his 'rational generalization' as 'complete' and 'exhaustive', terms which Kant applied to his own work as well.[14] Has Bohr been properly understood? Is his contribution as significant as he believed it was?

Let us look again at each of his central arguments. First, there is (B1): 'Some kind of conceptual framework is a necessary condition of the possibility of ordering experience.' Bohr assumes that our awareness of having experiences will not be challenged. But what constitutes this ability to have and to order experiences? Having accepted that physics works, he then reflects on the necessary conditions which operate in the augmenting and ordering of experience (the last words being roughly equivalent to his definition of physics). His usual overt defence of this statement (B1) begins by an appeal to the work of Galileo, Newton, and Einstein, whose new theoretical frameworks made a more general ordering of experience possible.

This first fundamental proposition also asserts not only that new sets of experiences may demand new frameworks, but also that conceptual frameworks of one kind or another are indispensable to the ordering of any set of experiences. Experiences, Bohr seems to be saying, are of themselves particular. He is not so much referring to the unity within any one immediate set of experiences as to the ordering of quite disjunct experiences. Thus, while it may be plausible to argue that the ordering of the colours and sounds and smells that make up my experience of a sunset over the harbour is due to the order that exists in the source of the experiences, I might also ask how I should correlate my various experiences of the motion of the sun in relation to the earth throughout the days of the seasons of the year. Without conceptual frameworks they will remain random and, for the one experiencing, disconnected. On the other hand, however, a notion of the sun waking and sleeping, or of it turning around the earth, or of the earth turning around the sun: all these are different frameworks for ordering such experience.

What is it, then, that patterns our experiences? If by definition two experiences are not one experience but separate, how can we link such experiences together except by some sort of provision of a connection on our own part? Remember that by 'conceptual framework' Bohr meant 'the unambiguous logical representation of relations between

[13] Bohr, *AHQP* interview 5, p. 3 of transcript.

[14] See section 3.6. above, and *Critique of Pure Reason*, A xiv.

experiences' (though one suspects that by 'logical' he also meant 'intelligible' or 'reasonable' as well as logically consistent). Bohr has in mind frameworks like space-time and causality and, more general still, complementarity.

This outline of an argument can be filled out through an appeal to Strawson's discussion of the re-identification of particulars. By 'particulars', as opposed to qualities and properties, Strawson means material bodies, historical occurrences, persons and their shadows: the sorts of things we experience rather than theorize about. One condition for identification of particulars is simple demonstration through hearing, touch, or other sensible indication. But when the particular to be identified is beyond the range of demonstrative identification, how can one avoid the possibility of reduplication, of identifying a different but otherwise identical particular?

Strawson suggests that it is plausible for non-demonstrative identification to rest securely on demonstrative identification: 'For by demonstrative identification we can determine a common reference point and common axes of spatial direction; and with these at our disposal we have also the theoretic possibility of a description of every other particular in space and time as uniquely related to our reference point.'[15] Though I have not been to the North Pole or climbed Ayer's Rock, I am able to refer to both of these places through the use of spatial references and other more immediate experiences such as using a compass or viewing a monolith. It is constitutive of our experience, in other words, that it occurs 'somewhere' and 'sometime', whether that space and time are story-relative and indirectly demonstrable, or immediately demonstrable. As far as Strawson is concerned, the conceptual system of spatial and temporal relations is part of the core of human thinking and human conceptual equipment.

Strawson connects this argument to the primary place of persons: 'for each person there is one body which occupies a certain *causal* position in relation to that person's perceptual experience'.[16] A similar, explicitly transcendental argument has been elaborated by Charles Taylor in his defence of the proposition that our perception of the world is that of an embodied agent. One of the first steps in Taylor's argument is to claim that our perceptual field (the collection of experiences at one place and time) 'must' have a structure. If there is no

[15] Strawson, *Individuals*, p. 22.
[16] Ibid., p. 92.

structure, we have no grasp of the thread of the world and are left instead with a confused debris of disconnected experiences.[17]

Bohr's argument is not quite so specific. He is not at this point stipulating that a space-time framework is essential to the ordering of experience, though he will have something to say about that implicitly in his next propositions. He is certainly not declaring that such a framework is universal and a priori necessary. If I as a subject have a set of experiences, he seems to say, then whatever puts those experiences into some sort of order will be the result of my own conceptual activity, whether learnt or created, and of my attempts to communicate that result to others with similar experiences and similar, if not identical, conceptual abilities. The argument is almost tautological, for Bohr defined a conceptual framework as an unambiguous logical representation of relations between experiences. Placing this definition in (B1), one obtains the following proposition: 'Some kind of unambiguous logical representation of relations between experiences is a necessary condition of the possibility of ordering experiences.'

Like all transcendental arguments, this first proposition simply has the character of laying bare the general features of our use of concepts. It is not the sort of argument that one can understand by considering a move from premiss to conclusion. Given the possibility that we are able to order distinct experiences, we must ask ourselves what necessary conditions constitute such an ability. We have to get the point for ourselves.

(B2) is a more elaborate claim: 'It is a necessary condition of the possibility of objective description of processes at the boundaries of human experience that concepts related to more normal experience be employed.' Here Bohr moves further into considerations about 'the actual structure of our thought about the world', the provenance of Strawson's descriptive metaphysics. The review of Bohr's argument in Chapter 3 (section 3.5.) told us a little more about his notion of objectivity. Given his refusal to accept any absolute subject-object distinction, 'objectivity' for Bohr does not mean an accurate one-to-one correspondence between outer reality and inner representation. At the same time, scientific objectivity demands 'the attempt to attain to uniqueness by avoiding all reference to the perceiving subject'.[18] In this light, (B2) concerns itself with the priority of public reference to definitely identifiable particulars and well-entrenched conceptual

[17] Taylor, 'The Validity of Transcendental Arguments', pp. 153 ff.
[18] *ATDN*, pp. 96 f.; see section 5.1. above.

frameworks in any attempts to extend the range of our experiences and to shape our description of such experiences.

Our common instinct, of course, is that all language is built up in this way: we can only move into the unknown if we have some idea of where we are; or, our language and our comprehension begin with very simple particulars and move on from there into the more abstract. Any new conceptual frameworks which we invent or uncover are in part the work of our own imagining, but also in part demanding of connection with existing frameworks and sets of experiences. Bohr is unable to find sophisticated philosophical arguments to defend this instinct, nor does he deal with the various questions which arise as a result of his assertion.

Once again, though I do not intend completely to identify Bohr with Strawson's early arguments, it is convenient to turn to Strawson for a more detailed account of the dependence of some categories of particulars on other more basic categories. Strawson notes that there are some kinds of particulars which can only be identified to other persons in terms of more accessible particulars. Just as Bohr does, Strawson cites two cases as standing out here, the identification of 'private' experiences and the identification of particles in physics: 'Certain particles of physics might provide one set of examples. These are not in any sense private objects; but they are unobservable objects. We must regard it as in principle possible to make identifying references to such particulars, if not individually, at least in groups or collections; otherwise they forfeit their status as admitted particulars.'[19] Strawson places a great deal of stress on the existence of material bodies in space and time as the key to the ordinary referring powers of individual persons, who are also material bodies in space and time. This is the ordinary, everyday 'classical' framework which Bohr himself acknowledges as primary in any attempts at objective description.

For example, I can distinguish between my two identical ball-point pens because of their different spatial locations in relation to my relatively stable spatial framework of reference. The one in my pocket was given to me by a special friend. But if I put them in a shoe-box, close the box, and shake them around, then, unless I have made some surreptitious identifying marks on them, I lose the thread of the possibility of re-identifying the one given to me by a friend as distinct from the one I bought in a shop. In quantum physics, according to Bohr, such surreptitious marks (or hidden parameters) are prohibited

[19] Strawson, *Individuals*, p. 44.

by the way nature is. Each observation of quantum objects, therefore, is something like opening the shoe-box: on each occasion we return to our usual spatio-temporal framework of descriptions, but we lose the thread of continuity. Quantum objects, however, cannot even be thought of in the same way as ball-point pens. We cannot envisage the quantum object itself, for we can only observe its interaction with macroscopic systems within our own frames of reference.

The claim in (B2) can be restated as follows: particulars which can only be identified indirectly must be identified by referring to demonstrable particulars if the identification is to be unambiguous. Otherwise, as Bohr insisted, how would we know what we were talking about? This proposition is thus another indispensability claim. As Strawson states the case: 'we could not speak of, and hence identify, particulars of the more sophisticated type unless we could speak of, and hence identify, particulars of the less sophisticated type'.[20]

The instinctive response here is to suggest that it must be possible to create a new language which is appropriate to the context of the new physics. Such a language, however, would always be parasitic upon our given language; and the success of this language, in turn, would be conditioned by the primary importance of re-identifiable particulars. This connection between ordinary language and materiality was certainly of interest to Bohr, and will be taken up again shortly.

Quantum physics, however, introduced new factors into the consideration of fundamental reality. Both implicitly and essentially, according to Bohr, it challenged the notion of an absolute and abiding spatio-temporal causal framework and of nature as an external continuum. This conclusion, moreover, coincided with his long-held epistemological belief that subject and object were as much a non-dual whole as they were arbitrarily separable. And the result of this conviction, as we have seen, is his prescription of complementarity.

What Bohr offers, therefore, is a new and more general conceptual framework or rational generalization. This is the proposition expressed in (B3): 'Our position as observers in a domain of experience where unambiguous application of concepts depends essentially on the conditions of observation demands the use of complementary descriptions if the description is to be exhaustive.' Implicit here are several stages in Bohr's way of thinking.

First, from (B1) and (B2) he concludes that the ordering of experience requires some sort of conceptual framework, and, to put it in Strawson's

[20] Ibid., pp. 44 f.

terms, that the objective description of non-demonstrable particulars demands reference to demonstrable particulars. Secondly, he is referring to the case in which the usual space-time framework for ordering experience no longer directly applies. What is essential to the character of a demonstrable particular is that it is evidently public; its very materiality locates it in space and time. This is not to say that there is such a thing as an absolute space-time framework, but that demonstrable particulars are sensible objects which exist in a relationship in space and time to those who sense them and refer to them. Such particulars have the character of the objects of classical physics, since they are clearly separate from the observers who refer to them. The space-time framework, furthermore, prevents the possibility of duplication of particulars: even if two particulars are identical, they cannot occupy the same position in the space-time framework. That is part of the condition of their sensibility and materiality.

According to Strawson's argument, material bodies, in a broad sense of the word, secure us to one single, common, and continuously extendable space-time framework of reference which is a 'fundamental condition' of identifying reference. Furthermore: 'a condition, in turn, of the possession of a single, continuously usable framework of this kind, was the ability to re-identify at least some elements of the framework in spite of the discontinuities of observation'.[21] Quite unwittingly, Strawson is suggesting a link between the success of our ordinary language about the world and the spatio-temporal mechanical models of physical systems implicit in classical physics. Bohr arrives at much the same conclusion, but by a more explicit consideration of the conditions limiting classical physics.

However, at the bounds of our experience we are dealing with 'private' particulars and unobservable objects. Here the problem of avoiding duplication looms large, for unambiguous reference becomes much more questionable. While we have good reasons for accepting the continuity of demonstrable material objects in space and time, like trees and other such bodies, we have no grounds for claiming the continuity of the experienced properties of such 'objects', tempted though we may be to presume this. The conceptualization of such particulars has to be related to the unambiguous identification of demonstrable particulars. Or, in other words, reference has to be made to the circumstances in which such experiences occur.

[21] Ibid., pp. 54 f.

Putting more sceptical objections aside, the tree outside my window is either an oak or a eucalypt, either deciduous or non-deciduous, but not both an oak and a eucalypt at the same time and in the same place. For sub-quantum events, however, reference in terms of classical concepts can only be made to interactions between the macroscopic apparatus and the atomic object. The interaction between the atomic object and the classical apparatus produces unambiguous, demonstrable results. Because the apparatus-object interaction cannot be sharply divided and must be treated as an unanalysable whole, the identification of the quantum object as an independent particular is essentially impossible. This being so, it cannot be regarded as having a space-time path. On the other hand, if we do not interfere with the quantum object, then we can think of it as having a space-time path, but, until we set up some sort of quasi-causal interaction with macroscopic apparatus, we can say nothing about it, and therefore, according to Bohr, we should not even dare to think about it. In this sense, space-time and causality must be seen as complementary frameworks.

The important point here is that observations at the boundary of experience, where sharp delineation between observer and observed is impossible, disclose the limit to the coherent applicability of the continuous causal space-time framework. Yet such observations still leave us with sets of experiences which we try to order. It follows from (B1) and (B2) that we must on the one hand provide a new conceptual framework and that, on the other, we are restricted to using concepts which take their unambiguous meaning from the everyday world of demonstrable particulars. Though it may go against the grain of ordinary experience, if the continuity of the causal space-time framework no longer applies, then we are entitled to use mutually exclusive concepts to give a complete description of what has been observed.

This claim might be taken as a comment on the conditions applying to the completeness of predication, to use Strawson's terminology once more.[22] Strawson considers the problem of the relationship between a set of descriptions and a fact, dealing with such questions as: when is the set of descriptions complete? how many descriptions must be taken into account before the description matches the fact? Bohr is making a related but different point, and one which takes us into new territory. He is asking about the conditions which apply when our very observations of fact, and therefore our descriptions, are necessarily

[22] Ibid., pp. 191-4.

constrained by the manner of observation. For example, two people observing a third person may refer to either 'that man' or 'John'. Here the constraints on description are contingent on previous knowledge, or lack of it, of the third person. In Bohr's case, however, the observation-reports on an electron which state 'it is a wave' or 'it is a particle' are related to the necessary limits imposed by the quantum condition rather than by the contingent ignorance of the observer. Further refinement of the predication is out of the question, for in this case our knowledge is limited and our ignorance invincible. Given (B1) and (B2), the only course of action open to us is to accept the complementarity of apparently irreconcilable opposites.

As pointed out in Chapter 3, there is no real contradiction here. Concepts are taken from one framework and applied to another. This is not to create a schizoid identity for whatever is under observation, but to accept the necessity of contrasting descriptions in order completely and unambiguously to describe what is being observed. Bohr's non-dualism is not irrationally mystical, but radically reasonable. He does ask us, however, to go beyond our normal familiar instincts in the way we use language, instincts which seek greater completion in our predication. He leaves the world a little beyond our grasp, albeit within our reach. This is not because the world is distant from us, though, but because of the nature of our participation, or even indwelling, in it.

Bohr is not claiming that complementarity is a purely a priori necessary framework. If he thought this way, his metaphysics would tend to be more revisionary than descriptive. Rather, given the limitation to the causal space-time framework, complementarity provides the only possible alternative framework for exhaustive and complete descriptions. It is complete because it takes all possible experience into account, and it is exhaustive because no alternative is possible. Nature forces it on us in a twofold way: first through the fact of the quantum of action as a universal and elementary constituent of nature; and secondly through the nature of our consciousness of what is spatially distinct from us and the conditions that apply to unambiguous communication in our usage of concepts and frameworks.

This is perhaps the most difficult aspect of Bohr's vision. The source of much vexation in coming to understand his position is that there is so little to be understood. He makes a reasonable claim rather than elaborating an inexorable argument. His critics imply that something like a leap of faith is required if one is to enter into his camp. Such criticisms, while not always charitably couched, are not inaccurate.

Transcendental argument entails reflection on given capabilities and then the articulation of the necessary conditions of such capacities 'to which we can conceive no alternatives'.[23] The argument only works when we engage in it ourselves.

Taylor has shown that transcendental arguments 'articulate a grasp of the point of our activity which we cannot but have, and their formulations aspire to self-evidence'. He then continues with some observations on the consequences of such methods: 'We have to innovate in language, and bring the limit conditions to experience to clarity in formulations which open up a zone normally outside our range of thought and attention. And these formulations can distort. And the deeper we go, that is, the richer the description, the more a cavil can be raised.'[24] Bohr's prescription of complementarity certainly has to do with the 'limit conditions of experience' and innovation in language. His attempts to 'open up a zone normally outside our range of thought and attention' are appropriately described as transcendental and, even, metaphysical. If we do not concede that this is the character of his thought, then it will elude us forever. Alternative formulations, it need hardly be said, are always possible. But they must be of a kind to capture this central aspect of his vision.

7.3. *Bohr among the Metaphysicians*

Strawson considers Kant and Aristotle to have been the two most eminent descriptive metaphysicians in the history of western thought, although one must allow for the fact that the term 'descriptive metaphysics' is a very general one, and that no philosopher can wholly be categorized as such. Here I wish to make some equally very general comments about those philosophies with which Bohr's stance might be compared. If 'purely' metaphysical speculation is remote from experience, and if it attempts to establish theoretical foundations of an esoteric kind (and Kant and Aristotle are not beyond such temptations themselves), descriptive metaphysics admits to an inability to dissect utterly the anatomy of human experience and knowledge. The comments made here do no more than indicate the elements of moderation and the emphasis on experience amongst the kind of metaphysical company which Bohr keeps.

[23] See Wilkerson, 'Transcendental Arguments Revisited', p. 114.
[24] Taylor, 'The Validity of Transcendental Arguments', p. 165.

The similarities between Bohr and Kant have been pointed out frequently; they have been touched on in the early chapters of this book and will be referred to again. It is also instructive in this context to note the tenor of those references to Aristotle which have been made in major studies of Bohr's point of view. Feyerabend, commenting on the radically 'empiricistic' character of both Aristotelian physics and quantum theory, observes that the 'Copenhagen Interpretation of quantum mechanics which Bohr adopted . . . reintroduces "Aristotelian" features of wholeness, but with excellent arguments'.[25] Jammer also remarks that Bohr has drawn the Aristotelian notion of 'unanalyzable wholeness' back into physics: 'Although because of its teleological component Aristotle's notion of "wholeness" is not entirely identical with Bohr's, Aristotle's insistence that the behaviour of a particle element, such as the earth, "must not be considered in isolation . . ." may be regarded as an early analog to Bohr's position.'[26] With the stress being placed first on experience and then on unanalysable wholeness, two features of descriptive metaphysics are brought to the fore in these comparisons with Bohr.

Along with Kant, Aristotle provides a point of reference in the writings of two of Bohr's most prominent disciples, Heisenberg and Weizsäcker (although they have developed their own marked variations on Bohr's position and Bohr would be dismayed to think that he had any disciples at all). Both note the appropriateness of Aristotle's notion of *potency* in the quantum account of nature.[27] Rather than accept the view implicit in classical physics, which treats objective reality as consisting of isolated objects, they suggest a revival of the Aristotelian emphasis on the underlying unity of matter.

The One which gives unity to nature–*physis*–is seen by Weizsäcker as the key to the possibility of any practice of physics, what we have referred to as a secret of success. In words which echo Bohr's dictum about the mutuality of actor and spectator, Weizsäcker 'studies the unity of science as a function of human performance, i.e., of experience,

[25] P. K. Feyerabend, *Realism, Rationalism and Scientific Method* (Cambridge, Cambridge University Press, 1981), 231 n. 17; see also p. 329 n. 76, and p. 341.

[26] M. Jammer, *The Philosophy of Quantum Mechanics* (New York, Wiley, 1974), 199 n. 66.

[27] See C. F. von Weizsäcker, *The Unity of Nature* (New York, Farrar Strauss Giroux, 1980), 52 f., 130, 346-56, and Heisenberg in W. Pauli (ed.), *Niels Bohr and the Development of Physics* (London, Pergamon, 1955), 13, and *Physics and Philosophy* (New York, Harper, 1962), 160, 180 f. See also P. Heelan's comparison of Aristotle and Heisenberg in his *Quantum Mechanics and Objectivity* (The Hague, Nijhoff, 1965), 142-55.

which, admittedly, is in turn possible only insofar as nature can in fact be experienced; in other words . . . the unity of science only in the still unanalyzed context of the essential interrelation of experiencing man and experienced nature—i.e., of the sought-for unity'.[28] Heisenberg, appealing to the same Bohrian image, concludes his reflections on *Physics and Philosophy* with the hope that modern physics will provide a new 'unification', 'a new kind of balance between thought and deed, between activity and meditation'.[29]

These successors to Bohr seek to restate a philosophy which avoids both extreme empiricism and positivism on the one hand, and pure a priori metaphysics on the other. Bohr himself, appealing neither to Kant nor to Aristotle, draws a modicum of support from the philosophies of the East. As the centre-piece for his coat of arms he chose the Yin-Yang mandala with its composition of opposites in unity; neither dualist nor monist, the attached motto that 'Contraria sunt Complementa' echoes the acceptance of paradoxical unity-in-difference which is to be found in the Vedas and Buddhist thought. Bohr once observed: 'For a parallel to the lesson of atomic theory regarding the limited applicability of such customary idealisations [as the distinction between object and observer], we must in fact turn to . . . that kind of epistemological problem with which already thinkers like Buddha and Lao Tse have been confronted, when trying to harmonize our position as spectators and actors in the great drama of existence.'[30] In Zen thought this search for the secret harmony is called the *Tao*, the 'Way' or the 'One'.

The *Tao te Ching*, attributed to Lao Tse, does not in fact contain any metaphysical argument as such, but its assertions on unity-in-difference are many. The way of paradox prescribed there is redolent with anticipations of Bohr. For example:

> The way that can be spoken of
> Is not the constant way;
> The name that can be named
> Is not the constant name. . . .
>
> Thus Something and Nothing produce each other;
> The difficult and the easy complement each other;
> The long and the short off-set each other. . . .[31]

[28] Weizsäcker, *The Unity of Nature*, p. 6.
[29] Heisenberg, *Physics and Philosophy*, p. 206.
[30] *APHK*, pp. 19 f.
[31] *Tao te Ching*, translated by D. C. Lau, I : 1-2, II : 39.

There is no evidence that Bohr ever read any of these works, but texts like the above are typical and may well lie at the root of his references. Bohr himself, I rather think, would also like to see reasons given for such assertions. We can only surmise.

Buddhist thinkers are more elaborate and varied in their emphases. The more abiding Buddhist philosophy, resting on the *via negativa*, is neither monist nor nihilist, but 'non-dualist'. Unity and difference are both acknowledged, as also is their mutuality. According to this interpretation of the vision of the 'enlightened one', as Alan Watts puts it,

> his point of view is not monistic. He does not think that all things are in reality One because, concretely speaking, there never were any 'things' to be considered One. For this reason both Hindus and Buddhists prefer to speak of reality as 'non-dual' rather than 'one', since the concept of one must always be in relation to that of many. The doctrine of *maya* is therefore a doctrine of relativity. It is saying that things, facts, and events are delineated, not by nature, but by human description, and that the way in which we describe (or divide) them is relative to our varying points of view.[32]

Bohr could not have subscribed entirely to such a view, for it entails a thoroughgoing relativism which leaves no place for recognition of the abiding harmony of nature. In this regard, such Buddhist thought is anti-realist. But if the philosophy here does not match his own conclusions, it suggests arguments similar to his own. Indeed, the word *maya* is derived from the Sanskrit root matr-, from which words like 'metre' and 'matter' are derived. The doctrine of *maya* has to do with measurement and the problem of observation, with material bodies and conceptual frameworks.[33] The Buddhist metaphysics of non-dualism at least arises out of the same issue to which Bohr gave so much attention. When Bohr explicitly appeals to the 'kind of epistemological arguments with which . . . Buddha and Lao Tse have been confronted', it must be these issues that he has in mind.

To many European thinkers the wisdom of the East seems more mystical than metaphysical, more religious than reasoned. And yet both traditions share some common ground as well as displaying some common patterns of thought. Mysticism seeks the One directly through experience whereas metaphysics follows the way of argument and reflection, but both aim at a common, if distant, goal, the grasp of the

32 A. Watts, *The Way of Zen* (Harmondsworth, Penguin, 1972), 60.

33 Ibid., pp. 59-63, and C. Humphreys, *Buddhism* (Harmondsworth, Penguin, 1967), 150 f.

underlying harmony in the manifold of our experience. A similar convergence has been identified in Bohr's thought: on the one hand there is his explicit account of the epistemological lesson which quantum physics has forced us to reconsider; and on the other hand, though less unequivocally stated, there is his attitude towards an infinite harmony of nature and a mutuality of spirit and matter, mysticism and rationality, noted in his remarks on Newton and his correspondence with Pauli.

Kant and Aristotle could not readily be described as seeking the One via a mystical path. What characterizes their more rigorous metaphysics is not only the respect which they pay to experience, but also their sustained effort to assert and explore the reciprocity of thought and sensibility and of spirit and matter. More patently than Kant, Aristotle assiduously steers a path through the extremes of idealism and materialism. Indeed, to a thoroughgoing holist like Rorty, Aristotle is far more of an ally than an enemy.[34] When confronted with the problem of subjectivity and objectivity, of One and Many, both Aristotle and Kant shift the argument away from Parmenides' abstract ontological grounds to the more accessible territory of human experience.

However, there is more of an assumption than a denial about the existence of the experienced external world, and Kant and Aristotle focus on the dynamics of our activity in the world and the conditions which constitute our abilities. In different ways they play both ends against the middle in giving priority to experience of the world and yet claiming a controlling role for the exercise of human reason.

Aristotle is just as interested in 'coming to be' as he is in 'being'. His attention is centred on the mutuality of potency and act, substance and accidents, essence and existence. Kant adopts a different vocabulary and, to his mind, vastly different conclusions. His emphasis on reason appears to many to be an idealism and a disparagement of the reality of the world of our experience. It can be argued, however, that his position includes a moderate realism.[35] As MacKinnon notes, both Kant and Aristotle 'accepted the physics that worked and then revised their norms of intelligibility'.[36] Bohr perhaps moved in the opposite

[34] See R. Rorty, *Philosophy and the Mirror of Nature* (Oxford, Blackwell, 1980), 40, 45.

[35] See, for example, K. Ameriks, *Kant's Theory of Mind* (New York, Oxford University Press, 1982).

[36] E. MacKinnon, *Scientific Explanation and Atomic Physics* (Chicago, University of Chicago Press, 1982), 370.

direction, but I would also symphathize with the view that his physics and his philosophy are indistinct from each other.

I am not suggesting any concrete similarities between Bohr and these two eminent descriptive metaphysicians, nor am I claiming that Bohr's sketchy philosophy is on the same level as their masterpieces. But their outlooks and the contours of their works do have common features, none the less.

One final question here: Why did Bohr refuse to appeal to the arguments of his apparent precursors? Had he misconstrued their purpose, or was he unfamiliar with their thought? He was generally unsympathetic towards Western philosophy, or what he knew of it. MacKinnon notes that Bohr never refers to any philosophers, and suggests that he 'tried to get others to think as he thought'.[37] Perhaps the answer to such a question can also be found within the very methods adopted: descriptive metaphysics rests not so much on abiding syllogisms as on reflection upon one's own capacities in language, experience, and knowing. Bohr's concepts were not likely to be those of ancient Greece or modern Königsberg. Kant's categories were not Aristotle's. Kant's doctrine of the synthetic a priori was not Aristotle's hylomorphism. Though their resolutions entail a principle of complementarity which ranges across the field of metaphysical issues, Bohr's solution was always going to be expressed within the context of his own set of conceptual frameworks and physics.

One could complain that Bohr leaves us sitting on the fence on many philosophical issues. It is true that, in his ignorance of the usual philosophical boundaries, he walks with carefree steps through the minefields of those traditional arguments which separate, say, empiricists and idealists. And he does indeed occupy the middle ground between extreme positions. He even seems to tell professional philosophers that 'both-and' is preferable to 'either-or'. His amateurish instincts, however, are closely tied to the way physics works; and that, as it were, is that, for Bohr was no amateur when it came to how physics worked.

In firmly rejecting the absoluteness of subject-object distinctions, analytic-synthetic distinctions, and word-world distinctions, Bohr also closes off much philosophical debate about mind, truth, and reality. But it is wrong to say that he turns his back on such questions, for he provides an alternative way of proceeding not only in science, but also in psychology, theology, anthropology, and so on. The quantum of

[37] Ibid., p. 376.

action, emblem of fundamental discontinuity in nature and of boundaries to our normal experience, remains his warrant in this.

If the quantum is to remain at the heart of physics, as Bohr argues against Einstein, then something of the permanence of scientific discovery is found to be congruent with the usually fluctuating fortunes of epistemology and metaphysics. Because Bohr's epistemology is tied to the way the physics works, it is a descriptive metaphysics offering a rational generalization and an all-embracing point of view. Bohr is well aware of the convergence of philosophy and physics in his own work: 'In our own century the immense progress of the sciences has . . . given us an unsuspected lesson about our position as observers of that nature of which we are part ourselves. Far from implying a schism between humanism and physical science, this development entails a message of importance for our attitude to common human problems, which—as I shall try to show—has given the old question of the unity of knowledge new perspective.'[38] Again, in his introduction to *Atomic Physics and Human Knowledge*, Bohr places this conviction in the very first paragraph: 'the study of the atomic constitution of matter . . . has thrown new light on the demands on scientific explanation incorporated in traditional philosophy'.[39] The way to scientific explanation is no longer to be guided solely by the paradigm of a continuous space-time framework, but also by the acceptance of fundamental limitations to the proper application of such a framework.

The reconciliation of philosophy and physics is not just a possible consequence of Bohr's work, for his epistemology also anticipates his physics. He is neither an experimental physicist nor an a priori metaphysician. His reflections on the conditions applying to unambiguous communication, and in particular on the relationship between conceptual frameworks and limit-experiences, belong to that enterprise which Strawson designates as descriptive metaphysics.

Can anything more be said about a purely philosophical defence of Bohr's position? Is there a short cut for the non-physicist, unwittingly caught in a classical empirical world-view? Holistic explanations notoriously defy further reduction.[40] The holist account of knowledge is an alternative to foundational epistemology, and the two schemes are not readily reconciled. Precisely because it is descriptive rather than revisionary, Bohr's argument is circular. Yet, as Heidegger has argued

[38] *Essays*, p. 8.
[39] *APHK*, p. 1; see also p. 72.
[40] See C. Peacocke, *Holistic Explanation* (Oxford, Clarendon, 1979), 18-41, 212-16.

in his defence of our need to enter into this 'hermeneutic circle', the exploration of our 'relatedness backward or forward', between our being and our being in the world, confirms rather than mocks our capacity for making meaning.[41]

Bohr's vision should not be examined as either purely philosophical or purely physical. These are aspects of his thinking rather than separate components. I suspect that what frustrated Bohr in his disappointment at the response of professional philosophers (presumably meaning the positivists who would claim him as their own), was their inability to grasp a more open view of our participation in nature and our description of it. Quantum physics may well alter philosophy and society just as much as classical science did in the Enlightenment. The framework of complementarity, once we have become comfortable with it, may prove as radical an extension of our outlook as the shift from flat-earth theory to the understanding of the earth as a spherical planet. The union of microphysics and metaphysics in Bohr's vision will, in that case, remain 'something of very great importance'.

7.4. *Conclusion: Language and Mechanics*

Associated with Bohr's fundamental arguments in his interpretation of the quantum formalism is a provocative hint at *a link between the given character of ordinary language and a deterministic-mechanistic view of the workings of nature.* In other words, Bohr the philosopher-physicist works with the instinct that classical physics is the inexorable result of the use of language based on the identification of experienced material particulars; or, vice versa, the use of a language based on identification of experienced particulars will ultimately lead to a sense of the persisting presence and movement of material objects in space and time, and hence to principles of conservation, causal change, and continuous space-time frameworks. The longer one spends pondering what it was that Bohr wanted to communicate, the more this particular aspect of his thinking attracts attention.

We have said enough in the preceding chapters about Bohr's views that we are 'suspended in language' in our attempts to offer unambiguous descriptions of what we experience and observe. This is certainly a key element in Bohr's outlook. All sophisticated theoretical vocabulary and all mathematical formalisms ultimately must in part be reduced to

[41] See M. Heidegger, *Being and Time* (Oxford, Blackwell, 1976), 7 ff., 153, 315.

ordinary language for their unambiguous meaning. Perhaps enough has not been said, however, about Bohr's views on the mutuality of space-time frameworks and the conservation of energy and momentum. Crucial to a determinist mechanics, of course, is not only a space-time causal account, but also the principle of the conservation of energy and momentum in any exchanges of energy and momentum. Without such a rule it would be impossible to predict future states of interacting systems. The quantum of action, however, marks the renunciation of this 'action principle', which 'symbolizes, as it were, the peculiar reciprocal symmetry relation between the space-time description and the laws of conservation of energy and momentum'.[42]

What is especially remarkable about Bohr's vision is that he commonly links together these apparently quite separate lines of thought concerning ordinary language and mechanist frameworks. Immediately prior to and immediately following the quotation just given, for example, he discusses the problems of observation and description and the necessity of employing classical concepts. Ordinary language, he seems to say, is tied up with our customary physical ideas, and this is particularly so with respect to our ideas of conservation in space and time. Indeed, Bohr seems to state, our customary experience of conservation in space and time provides us with the framework for unambiguous reference. This is almost the reverse of Strawson's argument, which moves from the possibility of re-identifying material particulars to the necessity of space-time frameworks and the primitive notions of bodies and persons. Bohr puts his case this way: 'On the one hand, we must bear in mind that the meaning of these [fundamental physical] concepts is wholly tied up with customary physical ideas. Thus, any reference to space-time relationships presupposes the permanence of the elementary particles, just as the laws of conservation of energy and momentum form the basis of the concepts of energy and momentum.'[43] Bohr's chief concern in passages such as this is to explain the limits to the applicability of the classical determinist account of mechanics. Between the lines, however, there remains an intriguing insight into a connection between the basic elements of our language and notions of space-time and conservation.

The determinist-mechanist account of nature can be crudely compared with a billiard-table model of colliding and rebounding spheres. It may be debatable whether or not any classical physicists actually held such a

[42] *ATDN*, p. 94.
[43] *ATDN*, pp. 113 f. See also p. 94, *APHK*, pp. 72, 90, and *Essays*, pp. 5, 11.

Laplacean view in its entirety, but this is the ideal of classical physics with which Bohr takes issue. Crucial to the success of the mechanist model is the assumption that the physical objects are entirely independent of the effects of observation, so that any measurements of the state of the system at any point in space and time do not interfere with its future pathways in space and time. A second assumption is that these independent physical bodies are permanent and that they maintain the properties we observe in them whether we are observing them or not. Hence concepts like inertial state and conserved momentum reflect our belief that independent physical bodies maintain the physical properties we observe in them whether we are directly referring to them or not.

Bohr points out these assumptions frequently enough, just as he uses the discovery of the quantum discontinuity to demonstrate their limitations. He also takes this discovery as an illustration of the need in all spheres of knowing to take account of the circumstances of observation: 'The gist of the argument is that for objective description and harmonious comprehension it is necessary in almost every field of knowledge to pay attention to the circumstances under which evidence is obtained.'[44]

Such a conclusion, however, is not meant to leave us in a hapless relativism. Bohr does want to defend a kind of objectivity, and he does want to claim that language can be unambiguous. It is at this point that the intriguing and covert connection between ordinary language and mechanistic descriptive concepts come to the fore. Using something like Strawson's arguments, Bohr maintains that language works because it is founded on the regularity of re-identifiable particulars in the sensible world. Such sensible particulars are clearly subject-independent, neither private nor mediated. Bohr is suggesting that there is a trap entailed in the success of language: we may be tempted into thinking that ideally all language must operate in the same way, so that derived or analogous application of concepts are similarly univocal and similarly refer to independent and enduring objects. For his part, however, Bohr wants to point out the trap and formulate ways for moving beyond it. In so doing, he attempts to outline broader notions of objectivity and realism. It is also interesting in this context to note that Strawson's more recent work examines the interdependence of conceptual capacities and beliefs and concludes that we must recognize

[44] *APHK*, p. 2.

a certain ultimate relativity in our conception of the real.[45] Critical realism becomes what might be termed relative realism.

The world of independent identifiable particulars is indeed open to being construed as a primitive mechanist system. This is because its elements are relatively permanent and independent in space and time, and because they exhibit relationships in space and time of causal interaction or of stable distancing. There is a spatial connection between my shadow and my body, for example, and such a shadow may disappear when the sun is covered by clouds. The status of this ordinary sensible world is a conservative one: particulars remain the same unless some change is caused. Bohr does not deny such a stability: indeed, he accepts it as the source of the success of our language and as the anchorage for our belief in the objectivity of nature. The intriguing point he is making, however, is that this does not give us the right to project the quasi-mechanist view of reality entailed in our language beyond its appropriate application.

Bohr is making the same point when he refers to 'the mutually exclusive relationship which will always exist between the practical use of any word and any attempts at its strict definition', or when he states that 'the nature of our consciousness brings about a complementary relationship . . . between the analysis of a concept and its immediate application'.[46] Both the definition of words and the analysis of concepts entail the notion that words and concepts are discrete entities. On the other hand, our practical application of words entails a holism of the conscious subjects and the described objects.

Bohr's emphasis on our need to generalize the scope of the applicability of our ordinary pictorial concepts arises out of considerations like these. His analysis of our ordinary use of concepts is indeed an original contribution to descriptive metaphysics. As he himself declares, the recognition of the quantum of action, together with a consideration of the conditions for unambiguous communication, has issued a sweeping challenge to our customary ways of thinking:

> In our century, the study of the atomic constitution of matter has revealed an unsuspected limitation of the scope of classical physical ideas and has thrown a new light on the demands on scientific explanation incorporated in traditional philosophy. The revision of the foundation for the unambiguous application of our elementary concepts, necessary for comprehension of atomic phenomena, therefore has a bearing far beyond the special domain of physical science.[47]

[45] See P. F. Strawson, *Skepticism and Naturalism* (London, Methuen, 1985), 42 ff., 94. [46] *APHK*, p. 52, and *ATDN*, p. 20. [47] *APHK*, p. 1.

In a final assessment of the significance of Bohr's contribution to our ways of thinking, it is on this point that our attention must be focused.

We are left, also, with a sense of the complexity and newness of Bohr's account of the description of nature. This complexity was as obvious to his colleagues as it was to those who took issue with him. 'The Quantum Theoretical Walpurgis Night' is a scene from a parody of Faust presented at Bohr's Institute for Theoretical Physics in 1932. In this scene, 'The Lord' (alias Niels Bohr) is in conversation with Dau (the Russian physicist Lev Landau) who is bound and gagged. The scene goes like this:

THE LORD Keep quiet, Dau! . . . Now in effect,
The only theory that's correct,
Or to whose lure I can succumb
Is . . .
LANDAU Um! Um-um! Um-um! Um-um!
THE LORD Don't interrupt this colloquy!
I'll do the talking. Dau, you see,
The only proper rule of thumb
Is . . .
LANDAU Um! Um-um! Um-um! Um-um![48]

Bohr was forever relentlessly seeking after what he wanted to say, and, secondly, found great difficulty in communicating it to others.

Because Bohr's vision seems so profuse and unclear to many, it has been found not only difficult to grasp at times, but also impossible to refute. It is often a compliment to a piece of philosophy to be able to find out where it has gone wrong, for this is a sign that the philosopher in question has been using clear arguments. Bohr's transcendental argument, however, cannot be directly confronted, for it is not so much an argument as a statement about indispensable conditions. One can only either reject the method or suggest an alternative assertion. It seems that Bohr was continually trying to sharpen an articulation of what he perceived to be necessary conditions pertaining to quantum physics and to unambiguous communication. Philosophically, following a line of argument refined by Strawson, Bohr's case is reasonable. And, from the point of view of physics, his case has its firm warrant until either quantum theory is overthrown or radically altered; but Bohr would regard the quantum as a universal and unalterable element in nature.

[48] The full text, translated by Barbara Gamow, is reproduced in G. Gamow, *Thirty Years that Shook Physics* (New York, Anchor, 1966), 171-214.

Bohr was not interested solely in what could be said easily and clearly about trivial questions. He thought that a passion for precision and clarity would unnecessarily diminish the horizons of knowledge and inadequately account for our experience. At the conclusion of Chapter 1, Bohr's delight in a couplet from Schiller was noted: 'Nur die Fülle führt zur Klarheit, / Und im Abgrund wohnt die Wahrheit'. Real clarity was not to be found merely by careful analysis, but also by including the profusion of experience in a comprehensive framework; and real truth was not to be found in that which could be 'objectively' encompassed, but only by the recognition that all knowledge takes us to the edge of a precipice. Bohr's point of view, in other words, was incompatible with many contemporary Western ambitions in both physics and philosophy. And, furthermore, it is unfair to Bohr to judge his position from standpoints which he has moved well beyond.

Yet Bohr's descriptive metaphysics does shape old truths in new ways. He is adamant about the mutuality of subject and object and he is a kind of critical realist. Although he does not say so, his position implicitly depends a great deal on the role of what might be called reasonable and responsible judgement: his rejection of the absolute objectivity of a mechanist outlook goes hand in hand with a defence of what can be termed 'virtual' objectivity. Objectivity in knowing is not constituted by the precise equivalence of inner representations and an outer state of affairs, because no such strong subject-object distinction can be maintained, but rather by careful attention to the conditions for unambiguous communication about experience and the reasonable judgement that so-and-so is the case based on the success of ordinary language. 'In emphasizing the necessity in unambiguous communication of paying proper attention to the placing of the object-subject separation,' says Bohr, 'modern development of science has, however, created a new basis for the use of such words as knowledge and belief.'[49]

Bohr's realist discourse is two-tiered, with each tier being dependent on the other. There is the familiar reality of our ordinary discourse and experience. There is also the more subtle reality of quantum physics and of consciousness and emotions. His instinct, it seems, is that we have placed too much priority on the former, rather than recognizing the mutuality of both levels. His linguistic and holist account is more than purely pragmatic, though, for he holds that there is a coherence in nature to which our search for knowledge draws us asymptotically. In this fundamental vision, that we are not dreaming it all and that nature

[49] *APHK*, pp. 80 f.

is miraculously comprehensible, he is in agreement with Einstein. His defence of this point of view can be seen in his account of the eternal and infinite harmony of nature. If science is to work, as Dreyfus puts it, there must be a secret of success.

The overall force of Bohr's argument is that we are without absolute foundations in our participation in the world, despite the acceptance that our language works by being anchored in everyday experience of reality. Instead, we are agents operating within a given. It is in this quite simple if far-reaching sense that his approach can be said to be performative or self-referential. Such prescriptions may not be new—one hears in them echoes of Aquinas's insistence on the notion of *actus* and Heidegger's account of the hermeneutic circle—but they are salutary. If we start a journey in the wrong direction we may never arrive at the desired goal. So also, Bohr seems to say, philosophers have reduced elements of their interests into the wrong kind of distinctions. Such a judgement may be ill-informed and unfair, especially in today's climate of thought, but Bohr saw his own work as highly original and revolutionary.

The 'complementarity' description assumes that one is engaged in describing something as well as that there is something being described. The 'describing' suggests some sort of object of experience and description. Bohr also adds that the describing entails a 'describer' and subject who is just as much a part of the action as the object in question. If focus rests solely on the possibility of a detached object, then knowledge of reality is reduced to the analytic method, with its implications of materialism and mechanism, and neglects the whole. This, in turn, may give us a knowledge of the elements, but it fails to provide an account of the complete system.

Bohr thus pleads for analysis to be constantly balanced by synthesis, and for materialism never to be separated from its corollary in, as he put it, spiritualism. He is not referring here to ghosts dwelling dualistically in machines, but taking account of the subject, whether as a physical observing system or as a Cartesian *res cogitans*, and the infinite harmony of the 'whole' of nature.

At the core of Bohr's thinking is his *idée fixe*, complementarity. This framework for comprehension and synthesis is also a kind of master-metaphor, an emblem of pervasive force, an out-and-out attack on the subtle dualism that has pervaded Western mentality since the rise of classical science. The heritage of mechanist science has deeply influenced our individual and corporate outlooks. On the one hand we have the habit of placing souls in bodies like passengers in motor cars; on the other hand, from the perspective of mechanist materialism, we have done away with the passengers entirely. Bohr's account of unity-in-difference provides a challenge to both supernatural and mechanist accounts of reality. He is by no means alone in such an enterprise—one thinks of Bergson or Jung, for example, or Dostoievsky—but he makes his case from within the heart of physics and thus with a new and unique warrant.

Bohr's passion for poetry is not just an avenue for appealing comparisons, therefore, but runs parallel with his sense of words having metaphorical power as well as referring purposes. Words not only demonstrate obvious realities, but also invoke deeper realities. Even 'reality' was, for Bohr, merely a human word.[50]

The deep truths may not be easily comprehended, but they are the truths that matter most. Bohr's philosophical excursions are perhaps too brief to be regarded as a substantial work of descriptive metaphysics, but the avenue he directs us along is one of very great importance. Kalckar notes how Bohr used to say that since the distant days when human beings began to use words like 'here' and 'there' and 'before' and 'after', there had been no further advance in epistemology until Einstein's theory of relativity. It is a great challenge to our ways of thinking to accept the framework of complementarity as a broader generalization of the space-time causal framework which is tied so closely to our ordinary discourse. In the past we have witnessed the demise of flat-earth theories and earth-centred cosomologies. In the future, if Bohr is right, we must struggle to interiorize the implications of his interpretation of quantum theory and his conceptual analysis of our use of language.

Bohr's arguments, as I have identified them, cannot be shown to be logically inconsistent, as Einstein himself eventually admitted. This is partly due to the fact that they are transcendental arguments. We can only disagree with his insight or debate the articulation of that insight which he employs. This is not to say, however, that his work does not deserve further attention. In identifying his way of thinking and his chief concerns, I hope that I will have generated improved debate about Bohr's vision.

At the very end, having identified Bohr as a transcendental philosopher, a moderate holist, a kind of realist, and a descriptive metaphysican, *I wish now to remove such labels and remind the reader once*

[50] See J. Kalckar, 'Niels Bohr and his Youngest Disciples', in Rozental, *Niels Bohr*, p. 234, and Kalckar's book about his recollections of conversations about Goethe with Bohr, *Det Inkommensurable* (Copenhagen, Rhodos, 1985).

again that Bohr's thought is unique to Bohr. Many readers may feel that I have already stretched such labels to the point where they hardly carry any significance. Much of Bohr's descriptive metaphysics, for example, may also seem rather revisionary, for Bohr certainly seems to be offering a new structure for our thought about the world, even if he bases such a proposal in experience rather than in speculation. But I believe the various categories I have employed can be helpful in seeing where Bohr fits into the patterns of recent philosophy. And the difficulty in finding more accurate labels for Bohr also indicates how different and original his way of thinking is, even though his reflections are relatively direct comments on the connections between ordinary language and unambiguous reference.

Clearly, this study has been written by one who is a partisan of Bohr. Equally clearly, in conclusion, it suggests directions in which the exploration of Bohr's thought might move rather than pretending definitively to capture his vision. Finally, if this book is a testimony to a great mind, let it also be a reminder of Bohr's greatness of being. I conclude therefore with the eulogy offered by one of his youngest colleagues:

> In our memories Niels Bohr emerges as a picture of strength. His mind owned resources which never ran dry, and his nature held a radiance of life which gilded his personality with the gleam of eternal youth. No one who came into more intimate contact with him was the same when they left him as when they came. When I look back on the too few years during which I worked with Niels Bohr, they appear to me as filled with light and gladness. His easy laughter was hearty and infectious and bore a stamp of innocence which lightened one's heart. The assurance that Bohr's work will remain forever a living element in human thought cannot relieve the grief at parting.[51]

[51] Kalckar, 'Niels Bohr and his Youngest Disciples', p. 238.

Select Bibliography

I. *Works by Niels Bohr*

This bibliography does not include many of Bohr's more technical papers on physics. Note that most of the books are collections of articles, and the articles are also cited in section 2 below. The authoritative bibliography and collection of Bohr's writings is of course available in the many volumes of L. Rosenfeld, J. Rud Nielsen, and E. Rüdinger (eds.), *Niels Bohr Collected Works [NBCW]* (Amsterdam, North-Holland, 1972-).

1. Books (with abbreviations)

Atomic Physics and Human Knowledge [APHK] (New York, Wiley, 1958).

Atomic Theory and the Description of Nature [ATDN] (Cambridge, Cambridge University Press, 1934), English translation by J. Rud Nielsen and Urquhart of *Atomteori og Naturbeskrivelse* (Copenhagen, Copenhagen University, 1929).

Essays 1958-1962 on Atomic Physics and Human Knowledge [Essays] (London, Wiley, 1963).

On the Constitution of Atoms and Molecules: Papers of 1913 Reprinted from the Philosophical Magazine [OCAM] (New York, Benjamin, 1963), introduced by L. Rosenfeld.

The Theory of Spectra and Atomic Constitution [TSAC] (Cambridge, Cambridge University Press, 1922), English translation by A. D. Udden.

2. Articles (in chronological order)

'On the Constitution of Atoms and Molecules', i, ii, iii, *Philosophical Magazine* 26 (1913), 1-25, 476-502, 857-75 (reprinted in *OCAM*).

'The Spectrum of Helium and Hydrogen', *Nature* 92 (1913), 231 f.

'On the Spectra of Hydrogen', in *TSAC*, pp. 1-19 (English translation of 'Om brintspektret', lecture given in Copenhagen, Dec. 1913).

'Atomic Models and X-ray Spectra', *Nature* 92 (1914), 553 f.

'On the Effect of Electric and Magnetic Fields on Spectral Lines', *Philosophical Magazine* 27 (1914), 506-24.

'On the Series Spectrum of Hydrogen and the Structure of the Atom', *Philosophical Magazine* 29 (1915), 332-5.

'The Spectra of Hydrogen and Helium', *Nature* 95 (1915), 6 f.

'On the Quantum Theory of Radiation and the Structure of the Atom', *Philosophical Magazine* 30 (1915), 394-415.

'On the Series Spectra of the Elements', in *TSAC*, pp. 20-60 (English translation of 'Über die Serienspektren der Elemente', lecture given in Berlin, Apr. 1920).

'Atomic Structure', *Nature* 107 (1921), 104-7.

'The Structure of the Atom and the Physical and Chemical Properties of the Elements', in *TSAC*, pp. 61-126 (English translation of 'Atomernes bygning og stoffernes fysiske og kemiske egenskaber', lecture given in Copenhagen, Oct. 1921).

'The Structure of the Atom', *Nature* 112 (1923), 29-44 (English translation of Nobel Prize address, 'Atomernes bygning', given in Stockholm, Dec. 1922).

'On the Application of the Quantum Theory to the Structure of the Atom', *Proceedings of the Cambridge Philosophical Society Supplement* 22 (1924), 1-42 (English translation of 'Über die Anwendung der Quantentheorie auf den Atombau', *Zeitschrift für Physik* 13 (1923), 117-65).

'The Quantum Theory of Radiation', *Philosophical Magazine* 47 (1924), 785-802 (with H. A. Kramers and J. C. Slater).

'Über die Wirkung von Atomen bei Stössen', *Zeitschrift für Physik* 34 (1925), 142-57.

'Atomic Theory and Mechanics', *Nature* 116 (1925) 845-53, *ATDN*, pp. 25-51.

'Atomic Theory and Wave Mechanics', *Nature* 119 (1927), 262 (report of a lecture given in Copenhagen, Dec. 1926).

'The Quantum Postulate and the Recent Development of Atomic Theory' (one version of the lecture given at Como), *Nature* 121 (1928), 580-90, *ATDN*, pp. 52-91.

'The Quantum of Action and the Description of Nature', in *ATDN*, pp. 92-101 (English translation of the Planck Festschrift article, 'Wirkungsquantum und Naturbeschreibung', *Die Naturwissenschaften* 17 (1929), 483-6).

'Introductory Survey', in *ATDN*, pp. 1-24 (English translation of Introduction to *Atomteori og Naturbeskrivelse*, somewhat revised).

'The Atomic Theory and the Fundamental Principles Underlying the Description of Nature', in *ATDN*, pp. 102-19 (English translation of 'Atomteorien og grundprincipperne for naturbeskrivelsen', lecture given in Copenhagen, autumn 1929).

'Philosophical Aspects of Atomic Theory', *Nature* 125 (1930), 958 (report of lecture given in Edinburgh, May 1930).

'The Use of the Concepts of Space and Time in Atomic Theory', *Nature* 127 (1931), 43 (report of lecture given in Copenhagen, Oct. 1930).

'Chemistry and the Quantum Theory of Atomic Constitution', *Journal of the Chemical Society* 26 (1932), 349-84.

'Light and Life', *Nature* 131 (1933), 423-59, *APHK*, pp. 3-12 (English translation of 'Lys og Liv', lecture given in Copenhagen, Aug. 1932).

'Quantum Mechanics and Physical Reality', *Nature* 136 (1935), 65.

'Can Quantum-Mechanical Description of Physical Reality Be Considered Complete?', *The Physical Review* 48 (1935), 696-702.

'Causality and Complementarity', *Philosophy of Science* 4 (1937), 289-98 (lecture given in Copenhagen, June 1936).

'Biology and Atomic Physics', *Nature* 143 (1939), 268-72, *APHK*, pp. 13-22 (lecture given in Bologna, Oct. 1937).

'Natural Philosophy and Human Cultures', *Nature* 143 (1939), 268-72, *APHK*, pp. 23-31 (lecture given in Copenhagen, Aug. 1938).

'Analysis and Synthesis in Science', in *International Encyclopedia of Unified Science* I. 1 (Chicago, University of Chicago Press, 1938), 28.

'The Causality Problem in Atomic Physics', in *New Theories in Physics* (Paris, International Institute of Intellectual Collaboration, 1939), 11-30.

'Newton's Principles and Modern Atomic Mechanics', in *Royal Society Newton Tercentenary Celebrations* (Cambridge, Cambridge University Press, 1946), 56-61.

'Atomic Physics and International Cooperation', *Proceedings of the American Philosophical Society* 91 (1947), 137.

'On the Notions of Causality and Complementarity', *Dialectica* 2 (1948), 312-19.

'Discussion with Einstein on Epistemological Problems in Atomic Physics', in P. A. Schilpp (ed.), *Albert Einstein: Philosopher-Scientist* (Evanston, Northwestern University Press, 1949), 199-241, *APHK*, pp. 32-66.

'Atoms and Human Knowledge', in *APHK*, pp. 83-93 (English translation of 'Atomerne og den menneskelige erkendelse', lecture given in Copenhagen, Feb. 1949).

'Physical Science and the Study of Religions', in *Studia Orientalia Joanni Pedersen* (Copenhagen, Munksgaard, 1953), 385-90.

'Unity of Knowledge', in *APHK*, pp. 67-82 (lecture given in New York, Oct. 1954).

'Physical Science and the Problem of Life', in *APHK*, pp. 94-101 (revised version of a lecture given in Copenhagen, Feb. 1949).

'Quantum Physics and Philosophy, Causality and Complementarity', in R. Klibansky, *Philosophy in the Mid-century: A survey* (Florence, La Nuova Italia Editrice, 1958), 308-14, *Essays*, pp. 1-7.

'On Atoms and Human Knowledge', *Daedalus* 87 (1958), 164-74.

'The Unity of Human Knowledge', in *Essays*, pp. 8-16 (lecture given in Copenhagen, Aug. 1960).

'The Connection between the Sciences', in *Essays*, pp. 17-22 (lecture given in Copenhagen, Oct. 1960).

'The Genesis of Quantum Mechanics', in *Essays*, pp. 74-8 (English translation of 'Die Enstehung der Quantenmechanik', in F. Bopp, *Werner Heisenberg und die Physik unserer Zeit* (Braunschweig, Vieweg, 1961)).

'The Rutherford Memorial Lecture 1958: Reminiscences of the Founder of Nuclear Science and of some Developments Based on his Work', in *Essays*, pp. 30-73 (lecture revised and originally printed in *Proceedings of the Physical Society* 78 (1961), 1083-115).

'The Solvay Meetings and the Development of Quantum Physics', in *Essays*, pp. 79-100 (lecture given in Brussels, Oct. 1961).

'Light and Life Revisited', in *Essays*, pp. 23-9 (English translation of 'Licht und Leben—noch Einmal', lecture given in Cologne, June 1962; the unfinished manuscript being edited and translated not entirely by Bohr himself, and published posthumously).

II. *Other Works Cited*

Note that this is a very select bibliography, representing the more immediate interests of this study. For more complete surveys of the physics of the Bohr era, consult works such as those by M. Jammer or J. Mehra and H. Rechenberg.

AUDI, M., *The Interpretation of Quantum Physics* (Chicago, University of Chicago Press, 1973).

BASTIN, E. W. (ed.), *Quantum Theory and Beyond* (Cambridge, Cambridge University Press, 1971).

BELL, J. S., 'On the Einstein Podolsky Rosen Paradox', *Physics* 1 (1964), 195-200.

— 'On the Problem of Hidden Variables in Quantum Mechanics', *Review of Modern Physics* 38 (1966), 447-75.

BIERI, P., HORSTMANN, R., and KRÜGER, L. (eds.), *Transcendental Arguments and Science* (Dordrecht, Reidel, 1979).

— 'Das Adiabatenprinzip in der Quantenmechanik', *Zeitschrift für Physik* 40 (1926), 167-91.

BORN, M. (ed.), *The Born-Einstein Letters* (London, Macmillan, 1971).

BUNGE, M., (ed.), 'Strife about Complementarity', *British Journal for the Philosophy of Science* 6 (1955), 1-12.

— *Quantum Theory and Reality*, Studies in the Foundations, Methodology, and Philosophy of Science 2 (New York, Springer, 1967).

CLAUSER, J. F., and SHIMONY, A., 'Bell's Theorem: experimental tests and implications', *Reports on Progress in Physics* 41 (1978), 1881-1927.

de BROGLIE, L., 'A Tentative Theory of Light Quanta', *Philosophical Magazine* 47 (1924), 446-58.

d'ESPAGNAT, B., *New Perspectives in Physics* (New York, Basic Books, 1962).

— *Conceptual Foundations of Quantum Mechanics* (Reading, Benjamin, 1976).

DREYFUS, H. L., RORTY, R., and TAYLOR, C., 'A Discussion', *Review of Metaphysics* 34 (1980), 47-55.

DRIESCHNER, M., *Voraussage—Wahrscheinlichkeit—Objekt: Über die begrifflichen Grundlagen der Quantenmechanik* (Berlin, Springer, 1979).

EINSTEIN, A., 'Zur Theorie der Lichterzeugung und Lichtabsorption', *Annalen der Physik* 20 (1906), 199-206.

— 'Zur Quantentheorie der Strahlung', *Physikalische Zeitschrift* 18 (1917), 121.

— 'Physics and Reality', *Journal of the Franklin Institute* 221 (1936), 313-82.

— *Ideas and Opinions* (New York, Crown, 1954).

— Podolsky, B., and Rosen, N., 'Can Quantum-Mechanical Description of Physical Reality Be Considered Complete?', *The Physical Review* 47 (1935), 777-80.

Favrholdt, D., 'Niels Bohr and Danish Philosophy', *Danish Yearbook of Philosophy* 13 (1976), 206-20.

Feyerabend, P. K., 'Complementarity', *Aristotelian Society Supplement* 32 (1958), 75-104.

— *Realism, Rationalism and Scientific Method* (Cambridge, Cambridge University Press, 1981).

Folse, H. J., 'Kantian Aspects of Complementarity', *Kant-Studien* 69 (1978), 58-66.

— *The Philosophy of Niels Bohr: The Framework of Complementarity* (Amsterdam, North-Holland, 1985).

— 'Niels Bohr: Complementarity, and Realism', in A. Fine and P. Machamer, (eds.), *Proceedings of the 1986 Biennial Meeting of the Philosophy of Science Association*, vol. 1, (East Lansing, Michigan, Philosophy of Science Association, 1987), 96-104.

Frank, P., *Einstein—His Life and Times* (New York, Knopf, 1947).

French, A. P. (ed.), *Einstein: A Centenary Volume* (London, Heinemann, 1979).

— and Kennedy, P. J. (eds.), *Niels Bohr: A Centenary Volume* (Cambridge, Harvard University Press, 1985).

Gamow, G., *Thirty Years that Shook Physics* (New York, Anchor, 1966).

Heelan, P., *Quantum Mechanics and Objectivity* (The Hague, Nijhoff, 1965).

Heilbron, J. L., and Kuhn, T. S., 'The Genesis of the Bohr Atom', *Historical Studies in the Physical Sciences* 1 (1969), 211-90.

Heisenberg, W., 'Über quantentheoretische Umdeutung kinematischer und mechanischer Beziehungen', *Zeitschrift für Physik* 33 (1925), 879-93.

— 'Über den anschaulichen Inhalt der quanten-theoretischen Kinematik und Mechanik', *Zeitschrift für Physik* 43 (1927), 172-98.

— *The Physical Principles of the Quantum Theory* (Chicago, University of Chicago Press, 1930).

— 'Quantum Theory and its Interpretation', in S. Rozental (ed.), *Niels Bohr: His Life and Work* (Amsterdam, North-Holland, 1967).

— *Physics and Beyond* (London, Allen and Unwin, 1971).

Hermann, G., May, E., and Vogel, Th., *Die Bedeutung der modernen Physik für die Theorie der Erkenntnis* (Leipzig, Hirzel, 1937).

Hesse, M., *Revolutions and Reconstructions in the Philosophy of Science* (Brighton, Harvester, 1980).

Holton, G. J., 'The Roots of Complementarity', *Daedalus* 99b (1970), 1015-55.

— *The Scientific Imagination* (Cambridge, Cambridge University Press, 1978).

Honner, J. R., 'Niels Bohr and the Mysticism of Nature', *Zygon* 17 (1982), 243-53.

— 'The Transcendental Philosophy of Niels Bohr', *Studies in History and Philosophy of Science* 13 (1982), 1-29.

— 'On the Term "Transcendental" ', *Milltown Studies* 11 (1983), 1-24.

HOOKER, C. A., 'The Nature of Quantum-Mechanical Reality', in R. G. COLODNY (ed.), *Paradigms and Paradoxes* (Pittsburgh, University of Pittsburgh Series in the Philosophy of Science 5, 1972), 67-302.

JAKI, S. L., *The Origins of Science and the Science of its Origins* (Edinburgh, Scottish Academic Press, 1978).

— *The Road of Science and the Ways to God* (Edinburgh, Scottish Academic Press, 1978).

JAMMER, M., *The Conceptual Development of Quantum Mechanics* (New York, McGraw-Hill, 1966).

— *The Philosophy of Quantum Mechanics* (New York, Wiley, 1974).

JORDAN, P., *Der Naturwissenschaftler vor der religiösen Frage* (Oldenburg and Hamburg, Stalling, 1972).

KALCKAR, J., *Det Inkommensurable* (Copenhagen, Rhodos, 1985).

KLEIN, M. J., 'The First Phase of the Bohr-Einstein Dialogue', *Historical Studies in the Physical Sciences* 2 (1970), 1-39.

KRAMERS, H. A., 'The Quantum Theory of Dispersion', *Nature* 114 (1924), 310 f.

KUHN, T. S., *Black-Body Theory and the Quantum Discontinuity* (Oxford, Clarendon, 1978).

— Heilbron, J. L., Forman, P., and Allen, L., *Sources for History of Quantum Physics* (Philadelphia, The American Philosophical Society, 1967).

MACKINNON, E., 'Heisenberg, Models and the Rise of Matrix Mechanics', in *Historical Studies in the Physical Sciences* 8 (Philadelphia, University of Pennsylvania, 1979), 137-85.

— *Scientific Explanation and Atomic Physics* (Chicago, University of Chicago Press, 1982).

MEHRA, J. (ed.), *The Physicist's Conception of Nature* (Dordrecht, Reidel, 1973).

— and RECHENBERG, H., *The Historical Development of Quantum Theory*, vols. i-v (New York, Springer, 1982).

MEYER-ABICH, K. M., *Korrespondez, Individualität und Komplementarität* (Wiesbaden, Steiner, 1965).

MOORE, R., *Niels Bohr: The Man and the Scientist* (London, Hodder & Stoughton, 1967).

MURDOCH, D. R., 'Complementarity: A Study of Bohr's Philosophy of Quantum Physics', D.Phil. thesis (Oxford, 1981).

NIELSEN, J. Rud., 'Memories of Niels Bohr', *Physics Today* 16 (1963), 22-30.

PAIS, A., *'Subtle is the Lord . . .': The Science and Life of Albert Einstein* (Oxford, Oxford University Press, 1982).

PAPINEAU, D., *Theory and Meaning* (Oxford, Clarendon, 1979).

PAULI, W. (ed.), *Niels Bohr and the Development of Physics* (London, Pergamon, 1955).

PEACOCKE, C., *Holistic Explanation* (Oxford, Clarendon, 1979).

PETERSEN, A., 'The Philosophy of Niels Bohr', *Bulletin of the Atomic Scientists* 19 (1963), 8-14.

— *Quantum Theory and the Philosophical Tradition* (London, MIT Press, 1968).

PRZIBRAM, K. (ed.), *Letters on Wave Mechanics* (London, Vision, 1968).

ROBERTSON, P., *The Early Years* (Copenhagen, University of Copenhagen Academic Press, 1979).

RORTY, R., 'Verificationism and Transcendental Arguments', *Nous* 5 (1971), 3-14.

— *Philosophy and the Mirror of Nature* (Oxford, Blackwell, 1980).

ROSENFELD, L., 'The Epistemological Conflict between Einstein and Bohr', *Zeitschrift für Physik* 171 (1963), 242-5.

— 'Niels Bohr's Contribution to Epistemology', *Physics Today* 16 (1963) 10, 47-54.

ROZENTAL, S. (ed.), *Niels Bohr: His Life and Work* (Amsterdam, North-Holland, 1967).

SCHEIBE, E., *The Logical Analysis of Quantum Mechanics* (London, Pergamon, 1973).

SCHILPP, P. A. (ed.), *Albert Einstein: Philosopher-Scientist* (Library of Living Philosophers 7, Evanston Ill., Northwestern University Press, 1949).

SHIMONY, A., 'Physical and Philosophical Issues in the Bohr-Einstein Debate', preprint.

STAPP, H. P., 'The Copenhagen Interpretation', *American Journal of Physics* 40 (1972), 1098-116.

STOLZENBURG, K., *Die Entwicklung des Bohrschen Komplementaritätsgedankens in den Jahren 1924 bis 1929* (Stuttgart, Ph.D. thesis, 1977).

STRAWSON, P. F., *Individuals* (London, Methuen, 1966).

— *The Bounds of Sense* (London, Methuen, 1959).

— *Skepticism and Naturalism* (London, Methuen, 1985).

TAYLOR, C., 'The Validity of Transcendental Arguments', *Proceedings of the Aristotelian Society* 79 (1978-9), 151-65.

TOULMIN, S. (ed.), *Physical Reality* (London, Harper and Row, 1970).

von WEIZSÄCKER, C. F., 'Komplementarität und Logik', *Die Naturwissenschaften* 42 (1955), 521-9, 545-55.

— *The Unity of Nature* (New York, Farrar Strauss Giroux, 1980).

WILKERSON, T. E., 'Transcendental Arguments Revisited', *Kant-Studien* 66 (1975), 102-15.

Index